FROM HAND AX TO LASER

FROM

OTHER BOOKS BY
JOHN PURCELL

Flights to Glory

The Best-Kept Secret
(for young adults)

Class Report
(a novel)

NEW YORK

HAND AX TO LASER

MAN'S GROWING MASTERY OF ENERGY

By

John Purcell

Illustrated
by

Judy Skorpil

THE VANGUARD PRESS

ILLUSTRATION CREDITS

Except as noted below all the illustrations in this book are the work of Judy Skorpil. A number of them first appeared, in rather different form, in the pages of *Scientific American*.

Page 9: Academia Sinica

Page 21: King Features

Page 101: From CAUTIONARY VERSES by Hilaire Belloc, published 1941 by Alfred A. Knopf, Inc., reprinted by permission.

Page 121: Harper's Weekly

Page 269: Harper's Weekly

Page 271: *Scientific American* (weekly)

Page 287: Ben Rose

Published by Vanguard Press, Inc., 424 Madison Avenue, New York, N.Y. 10017.
Published simultaneously in Canada by Book Center, Inc., 1140 Beaulac Street, Montreal, Quebec H4R 1R8

Designer: Tom Torre Bevans.
Manufactured in the United States of America.
1 2 3 4 5 6 7 8 9 0

Library of Congress Cataloging in Publication Data

Purcell, John Francis, 1916–
From hand ax to laser.
Bibliography: p.
Includes index.
1. Power (Mechanics)—History. 2. Science—History. I. Title.
TJ163.9.P87 1982 621.042 81-16389
ISBN 0-8149-0860-8 AACR2

In memory of
Carleton Stevens Coon,
gifted scholar,
inspiring teacher,
valued friend

TABLE OF CONTENTS

FOREWORD

Our Energetic World

We seldom think about the different kinds of power we use or even about how we use them. We give even less thought to their origins. For example, if you live near the water, perhaps you went sailing last summer. You probably checked your watch to see if you had time enough or if the tides were right, drove down to the shore, took a dingy out to the sailboat mooring, climbed aboard, cast off, raised the sail, and took off. In the process you used no fewer than five different kinds of power. Can you count them?

Assuming that your watch has a mainspring rather than a battery, when you checked on the time you were taking advantage of a means of storing energy — the spring — that can be traced back to its simplest form, the bow, over a span of more than 10,000 years. The clockwork that keeps the spring from running down all at once has a shorter past, but it can also be measured in thousands of years.

The "self-moving" (auto-mobile) vehicle that took you to the shore has a mere century of history behind it, but the engine that makes its wheels turn harnesses the heat of internal combustion. The first internal-combustion engines were invented nearly 1,000 years ago; they burned gunpowder rather than gasoline and threw away their pistons each time they were fired. Today we call them cannons. The wheels you rolled on are not, of course, sources of energy in themselves, although many other kinds of wheels are used to harness power. Still, it is safe to say that wheels (and axles and cranks and all the other related gadgetry of rotary motion) have been one of man's most useful means of applying energy efficiently. And the first wheels are a good deal older than the 5,000-year-old Royal Tombs at Ur, where aristocracy's oldest wagons were found.

To return to the auto engine for a moment, the explosive gases in its cylinders were fired by an electric spark rather than by a gunpowder fuse. Making use of electric current for this purpose is the most up-to-date thing that has happened on your trip so far, although men were producing electric sparks long before Ben Franklin risked his life two centuries ago flying a kite in stormy weather.

If you rowed the dingy out to the sailboat mooring, you were using man's oldest power source: your own muscle. Moreover, other than the effort of walking, this was also your first serious application of manpower in your trip. Even so, you were not applying mere brute force. The oars that you pulled are ingenious multiplying devices — a short pull on the handle makes the oar blade travel a long way. An oar is a kind of lever, one of the five classes of "simple" machines that were recognized by the ancient Greeks more than 2,000 years ago.

Considered as a means of crossing water from place to place, your dingy has an even longer history than this. Until recently, the use of watercraft could not be traced back farther than to early Egypt and Mesopotamia, four or five thousand years ago. Studies of early Mediterranean trade in the kind of volcanic glass called obsidian now make it clear that men have been seafarers for twice that long.

Finally, you raise your sail and get under way. Assuming you were not caught in a flat calm, once the sail was raised you were using one of the two earliest sources of non-muscle power that man could exploit: the wind.

Stored power, the power of heat, electrical power, muscle power and wind power. How they do add up. Mankind has spent at least a million years learning to improve on the unaided power of his own muscles. His quest has met with steady success. In this book I will describe some of the most notable of these successes, beginning with man's emergence as a tool-using member of the mammalian order of primates. It will help, as a beginning, to describe man's unique place in nature; the starting line, so to speak, of the human race.

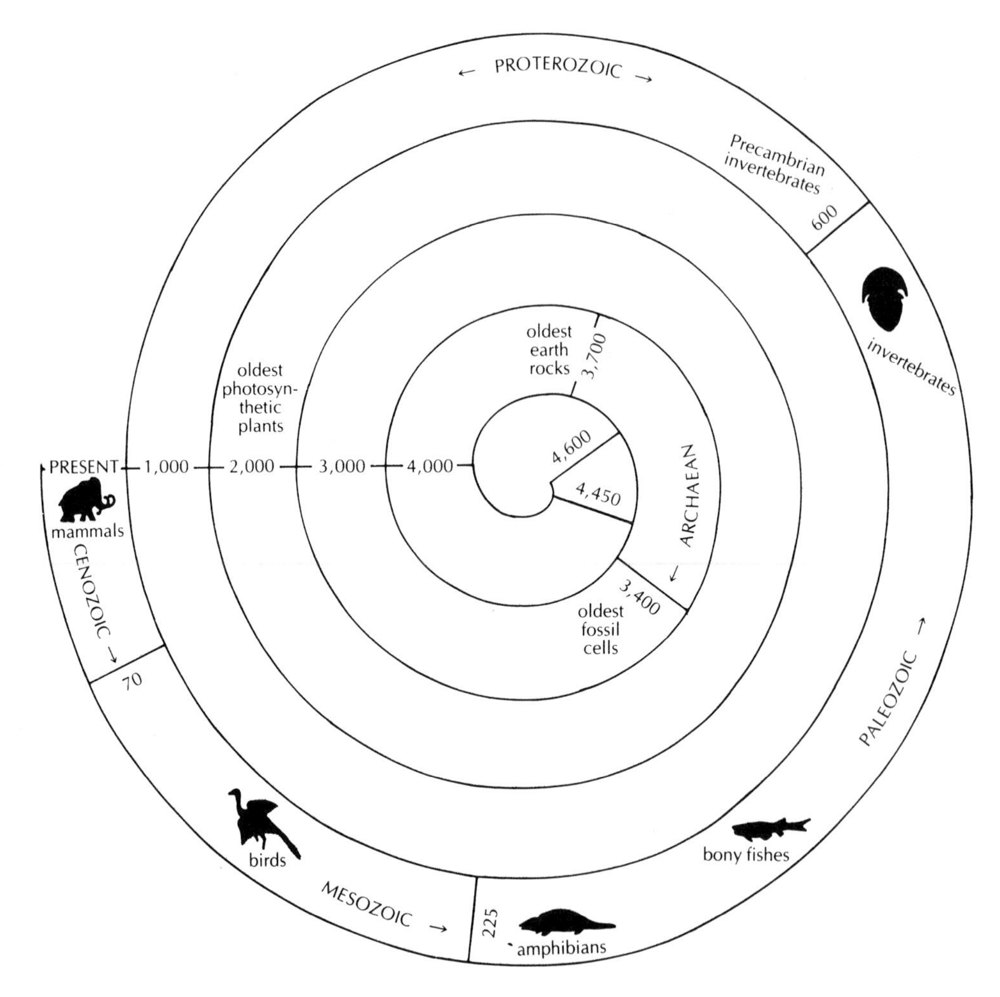

If we think of time as a spiral, like the outward-growing shell of a chambered nautilus, we can set the zero for the solar system at a point about 4,600 million years ago (*center*). We will also find that almost everything we know about life on earth belongs on the outermost turn of the spiral, from the Paleozoic (old life) Era, beginning 600 million years ago, through the Mesozoic (middle life) Era, beginning 225 million years ago, and the Cenozoic (recent life) Era, beginning 70 million years ago, to the present. The Cenozoic witnessed the rise of mammals; man's own rise has occupied less than 5 percent of Cenozoic time. (All figures are in millions of years.)

CHAPTER
1

Man's Place in Nature

ALL LIVING things, whether plants or animals, are marvelous individual packages of energy. The green plants capture the energy of sunlight; some of this energy is spent on growth and some is stored for later use. The animals that live by eating plants convert the plant tissue into muscle and fat. Other animals, who are not plant-eaters, accumulate their energy secondhand by eating the plant-eaters. Still other animals, such as bears and humans, take nourishment from plant and animal foods alike.

As is true of the plants, animals devote some of the energy they gain from food to growing larger. The rest of their daily energy budget is spent in other ways: in giving rise to new generations of their own kind, for example, and in making changes in the world around them. Within the animal kingdom, with its 14 great divisions, only the members of two divisions do much about changing the face of the land. These two are the arthropods — the "jointed legs" phylum that includes literally countless species of insects — and the chordates, the "rod-back" phylum that includes among the land-alterers some fishes and birds and several of our fellow mammals. Among the arthropods the social insects construct "houses" of a kind, where their young are fed until they reach maturity. A few fishes and most birds also build shelters, as simple as a scooped-out underwater hollow for fish eggs or as elaborate as the multiple nests where cliff-swallow offspring mature.

Among the mammals, a few invest energy in seeking out or digging dens and burrows or even, like the beaver, building "lodges." Man and a few of his primate relatives, however, make the principle efforts to construct shelters. Furthermore man is unique in the animal kingdom in locating independent energy sources and devising ways to harness them. Why should this be?

At least a part of the answer is "adaptability," and a great deal of man's adaptability stems from his biological heritage both as a mammal and, among the mammals, as the most successful of all the primates. Man's organic efficiency is a great help in his capture and use of energy. As a mammal, he is equipped with an internal thermostat that keeps his body temperature at an even level in hot and cold weather alike. As one of the advanced mammals he has passed beyond the egg-laying means of reproduction found among reptiles, birds and primitive mammals. Like the majority of mammals, women carry their young in embryo form within the body cavity until the offspring is mature enough to be born.

The efficiency of this means of bearing young, while often ignored, is easy to prove. For example, from any clutch of turtle eggs the hatchlings that manage to survive until maturity make up a very small fraction of the number of eggs laid by the mother. Many of the eggs are destroyed and many of the hatchlings are killed soon after they emerge. Mammals, on the contrary, suffer few losses between pregnancy and birth and fairly few between birth and maturity.

Consider a small population of mammals — 50 males and 50 females — with a far from outstanding reproductive rate. The 100 can be field mice, elephants, bats, dolphins, or anything you like. If in her entire life span each female among these 50 potential couples gives birth to only four offspring who reach maturity, and if the same is true of each successive generation, the population in the fifth generation will number 1,600 and, in the tenth generation, 51,000. Any biological factor that favors this kind of multiplication naturally increases the chances that the animal population in question will hold its own against competing spe-

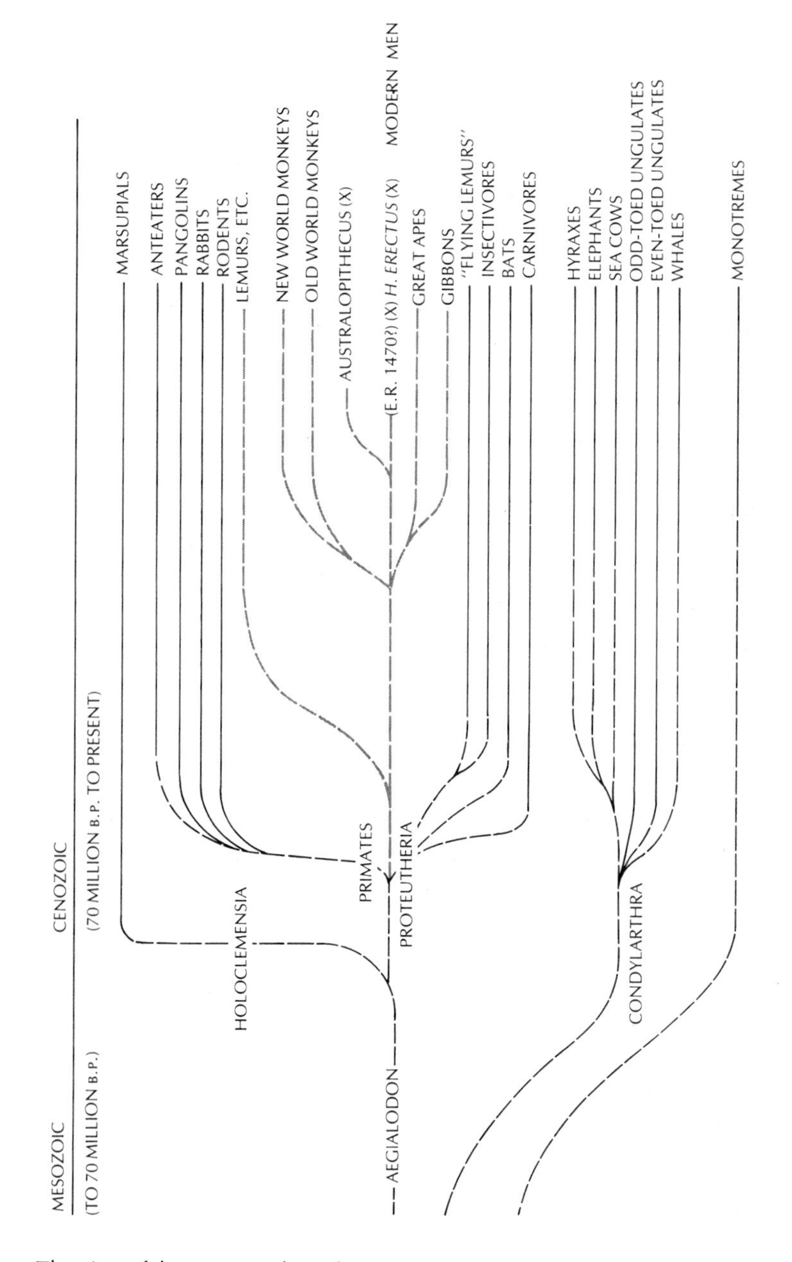

The rise of the mammals is shown schematically in this tree diagram; the order of primates has been placed centrally. An "x' indicates extinction, a fate that has befallen *Homo erectus,* his possible ancestor, "Ellesby" (ER 1470), and all species of *Australopithecus.*

cies. The mammals' special means of giving birth is just such a favorable biological factor.

So much for being a mammal. What are the biological benefits of being a primate? This is the order of mammals whose hands and feet have fingers and toes ending in nails rather than claws. Not only that, but the tip of the thumb can touch the tips of the other four fingers. These "prehensile" — that is, grasping — hands with their "opposable" fingertip-touching thumbs are one major benefit that man enjoys because he is a primate. Another benefit equally important but usually taken for granted, is a pair of eyes located on the front of the head in such a way that the two fields of forward vision overlap. This overlap is what gives man and other primates the ability to judge depth: to see things in three dimensions — from both sides at once, so to speak. Both these biological benefits would not amount to much, however, without a third. This is the ability, which is shared with many primates but which man alone has made his habitual posture, to stand upright.

Standing on one's own feet is another thing we take for granted, but if man still went on all fours — as his fellow primates who live on the ground do — he would not be a man at all. For it was standing up that first allowed man to use his grasping hands and distance-judging eyes to explore the world around him. The freeing of the hands for full-time manipulation — grasping, poking, and prying — guided in this exploration by eyes that see from both sides, is probably the single most important change in behavior in all of man's long evolutionary record. It is often said that what distinguishes man from his fellow animals is the fact that man uses tools. Certainly mankind's first tools were a pair of free hands, tools that were his because he stood upright.

There are a number of lesser biological benefits man has gained by being one of the primates. I have not yet mentioned, however, the greatest benefit of all. This is a brain that is quite large in proportion to overall body size. A large brain is not something that automatically appears along with a large body. Man's brain, on the average, makes up about 2 percent — one forty-fifth — of his average body weight. Some other primates have much more brain in proportion; for example, the weight of

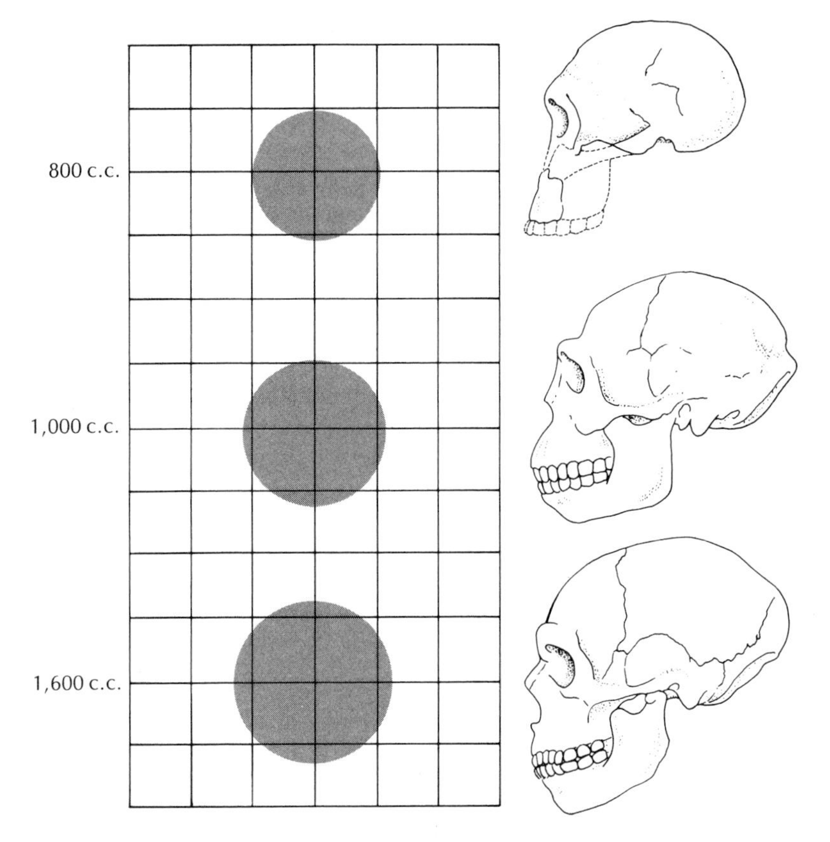

Three spheres (*left*) represent the increasing size of the brains of the three hominids whose skulls are shown (*right*). These are, at top, "Ellesby" (ER 1470), with a cranial capacity of 800 cubic centimeters; at center, a male specimen of *Homo erectus* from North China (1,000 cc); and, at bottom, a male specimen of *Homo sapiens neanderthalensis* from Europe (1,600 cc). Neanderthal brains were larger than today's average human brain.

the squirrel monkey's brain is almost 10 percent of its total body weight. Other primates have much less; the proportion in the sacred monkey of India, the macaque, is little better than one half of 1 percent. In both these examples, of course, the absolute size of the monkeys' brains is far below the human average of 1,360 cubic centimeters.

It should be said at once that it is not the size of the brain but its organization that is related to the rather misty concept we call human intelligence. An elephant's brain is much larger than a man's, and the brain of a great blue whale is far larger than an elephant's. Both, however, are much smaller in proportion to the animals' overall size than the brain of man is; the weight of the blue whale's brain is only one ten-thousandth of its body weight and the elephant's brain only one six-hundredth.

When volumes are given in cubic centimeters (cc for short) they are not easy to visualize. Let us, instead, assume that the various brains under discussion have been turned into perfect spheres rather than being their true shape, which is more like a very large half-walnut. If spherical, the average human brain (1,360 cc) would be a little more than five inches in diameter — about the size of a medium honeydew melon. The author of *Gulliver's Travels,* Jonathan Swift, had a 2,000 cc brain. As a sphere this would be a little over six inches in diameter, or about the size of a casaba melon. At the other end of the human scale, certain Australian aborigines have brains as small as 1,000 cc (a five-inch sphere, equivalent to a small honeydew), but show no signs of being subnormal because of this. Let us keep spheres like these in mind as we look at what the fossil record shows about the brain sizes of the near-men among whom the first humans arose.

It is still impossible today to trace the line of man's descent exactly through the fossil record. Compared to what the wisest men knew only 100 years ago, however, we have very much more evolutionary information today and can draw a broad general outline of the rise of man with confidence. For example, it is now clear that a not very apelike ancestor of man's ancestors, a primate known as *Ramapithecus* (that is, "Rama's ape"), could be found tens of millions of years ago not only in Asia and in Africa but also in eastern Europe.

So far, no skull of *Ramapithecus* has been discovered (although part of a face has been) and so the size of this primate's brain is unknown. In the case of a much later near-man, who first appeared on the scene some three million years ago, the fossil record is much richer. A number of skulls of *Australopithecus* (that is, "Southern ape") have been found and their brain size measured. The average is somewhat less than 500 cc. Transformed into a sphere, this would be equal to a medium, four-inch-diameter grapefruit, as compared to Swift's six-inch-plus casaba.

Australopithecus can best be described as an example of one evolutionary trend toward humanization. This near-man walked upright for much if not all of the time and so he shared with true man the empty hands that were the first tools. He may also have used chance objects as tools — fallen tree limbs, animal bones and horns, sharp-edged pebble fragments found in a stream bed. He lived in eastern Africa during an era of great changes in climate, and *Australopithecus* bones have been found from as far south as the Johannesburg area to as far north as Ethiopia. His time on earth lasted roughly from four million to around one million years ago, the same span of time that witnessed the rise of man.

Then *Australopithecus* died out. Could this be because his grapefruit-size brain was not able to sort out, store away, and draw conclusions from the thousands of messages sent to it each waking hour by his grasping hands and depth-perceiving eyes? Perhaps so, but even if true, the real cause may lie deeper. Perhaps his rather small brain lacked the kind of organization that characterizes the human brain, because 500 cc of gray matter is not quite enough to permit that kind of organization.

A few years ago, when no other candidates for the role of man's ancestor were on the scene except *Australopithecus*, the fact that this line of near-men had died out was a bitter pill for evolutionists to swallow. Indeed, a number of tries were made at branching true men off from the upper part of the *Australopithecus* tree. The need to do this came to an end a few years ago when other near-men — more advanced than any *Australopithecus* and equally as old as most of them — were found in Africa. Time may, in fact, prove these new discoveries to be man's immediate ancestors.

Today, the best-known of the newcomers has no identifying name but only a number: E.R. 1470. The two initials stand for East Rudolf, a region in the vicinity of Lake Rudolf (now named Lake Turkana) in Kenya where Richard Leakey and his associates found the fragments of a fossilized skull; 1470 is the catalogue number assigned to the reassembled specimen. Here I will call E.R. 1470 "Ellesby," both for convenience and in deference to Leakey's illustrious fossil-finding father, L. S. B. Leakey.

What has made Ellesby particularly important is that his brain size is nearly 800 cc, the size of an average four-and-one-half-inch cantaloupe. This is only about half an inch and 500-odd cc's away from modern man's medium honeydew. *If* the simple stone tools that Richard Leakey has found in the Rudolf region (although not in association with any fossils of near-men) could by some stretch of the imagination be taken for the handiwork of Ellesby and his fellows, we might here be seeing a borderline illustrated. Such a borderline in brain organization would then fall, in terms of size, somewhere between 500 and 800 cc. *Australopithecus* may have used the bones and pebbles that are associated with his remains as tools but no indisputable evidence exists that these objects were fashioned by him. The Lake Turkana tools, however, are certainly the products of willful manufacture rather than the result of natural breakage. Was the cantaloupe brain that much better organized than the grapefruit? Perhaps time will tell.

At least this much is certain even today: not much more than two million years later true men had arrived on the scene. They were present some 500,000 to 750,000 years ago in family-size if not clan-size numbers in northern China and their solitary remains have been unearthed elsewhere in China, in Java, in East Africa and (quite probably) in northern and southern Africa and in eastern and western Europe. Who were these true men?

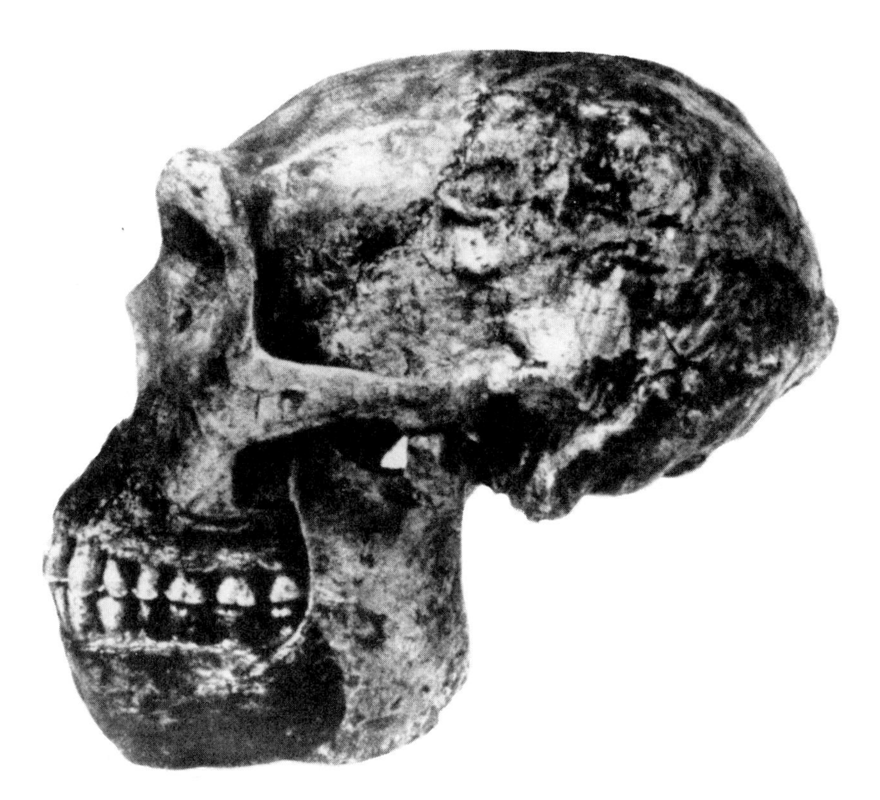

Where we can be sure we all started: this is a profile view of a typical *Homo erectus* skull. The first specimens of this early man were found in Java late in the 19th century and the discoverer, Eugène Dubois, gave the hitherto unknown primate the name *Pithecanthropus erectus,* or "apeman upright." When it was finally agreed that these early hominids belonged to our own genus, *Homo* replaced "apeman."

Today we call them *Homo erectus* (that is, "upright Man"). This is to say that they belong to the genus of true men, *Homo*, but not to our own species, *Homo sapiens* (that is, and please, no smirking, "wise Man"). Back in the good old days of fossil-hunting, when each discoverer felt free to name his finds whatever he chose, the first of these true men to be found, in Java, was christened *Pithecanthropus erectus* (that is, "upright Apeman"). The true men found later in China — at first only a few teeth but, after that, jaws and skulls and other bones — were seen to be related to the Java specimen, but they ended up with their own patriotic name anyway: *Sinanthropus pekinensis* (that is, "China man from Peking"). Equally patriotic names have been given to those found in other parts of the world, but the students of the species *Homo erectus* are inclined to ignore these formal christenings. In the past 50 years or so, fossil hunters have uncovered many more of these first true men. There are now no fewer than 7 from the fossil beds of Java, at least 2 from Olduvai Gorge in East Africa and more than 40 from China.

All of the skulls of these true men that have been made available for brain studies held brains that, with one exception, were as large or larger than Ellesby's. Three of them, in fact, equal or outrank those of living aborigines. This push beyond cantaloupe to near-honeydew ought to have been reflected in these true men's use of tools. As luck would have it, only the group living in caves near Peking have been found in undoubted association with campsite debris. The trash included many animal bones, mostly of deer, the seeds of wild berries, stone tools made out of coarse-grained quartz not found in the cave region (and therefore carried there from elsewhere) and other bits of bone and antler that appear to have been used as tools. Perhaps most important of all, the caves contained layers of ash, charcoal, and burned animal bones. The true men of northern China thus knew how to keep a fire burning, although they may not have known how to start one, and they had learned enough about the value of fires to take the trouble to tend them.

Making tools, eating venison, and sitting by the fire in a

cozy cave; an enormous gulf separates this scene from the earlier one wherein some as yet unknown near-men first habitually stood on their feet and began to use their liberated hands as tools. What allowed man to span that gulf was his biological heritage: his place in nature as an adaptable and efficient mammal of primate lineage. Not all mammals are family animals, but enough of them are to let us suggest that man's sociality also stems in part from his heritage as a mammal. Almost all primates are manipulators and so we can say that man's handiness is a direct inheritance from the primate line. As to the upright posture that left his hands free, this heritage he shared with one near-man that we know about, the now long-extinct *Australopithecus.*

Was it the increase in brain size — certainly a biological rather than a cultural development — that accounts for the success of man's precursors and the failure of *Australopithecus?* We have no clear answer, but only the strong suspicion that no harm comes from having a big brain. Is there an organizational borderline somewhere between 500 and 800 cc, or between 800 and 1,000 cc? Here we must leave the question to be answered by future students and for the present let the record itself speak. The grapefruit-brains became extinct. The honeydew-brains built enough on their biological heritage to end up in firelight, eating meat and making tools. Once having captured fire — the first kind of energy he exploited other than his own muscle — true man, as we shall see in the chapters ahead, has steadily enlarged his capacity to exploit and control many other sources of energy.

CHAPTER

2

What is Energy?

It is time to talk about the object of man's quest: the control of energy and its conversion into power. A good place to start is that firelit cave near which, a few hundred thousand years later, the great city of Peking would arise. By capturing fire, the people of Chou-kou-tien took possession of an artificial source of energy that was completely independent of their own muscle power. Why "artificial"? Because the cave folk could keep their fires alive only by the artifice of feeding them fuel. But where did the energy come from in the first place? The answer is: from the sun.

Trace the steps from the flaming center of our solar system to the glowing hearths in China. The fuel that kept the fires burning was wood. We know this is so because charcoal has been found in those hearths. We could safely have guessed this anyway because, with the exception of dung, oils, fats, and wax, the only fuels known to man until a few thousand years ago were plant products — and wood in particular. Indeed, until a few hundred years ago wood and the wood by-product, charcoal, were the only fuels in general use. So the first step in the transfer of solar energy from star to hearth was the application of a little human muscle power as the cave folk moved about collecting firewood. We can only assume they picked up fallen tree limbs and similar litter; their stone tools were not at all suited to chopping down trees.

The second and final step in the transfer you have probably already guessed. Trees have to grow before there can be tree limbs to fall — a growth made possible by the availability of solar energy at the surface of our planet. We shall not consider here the elaborate ins and outs of the photosynthetic mechanism that converts water and atmospheric carbon into food for plants. Lest you think, however, that the sun's only contribution to the growth of a tree is the tiny amount of radiant energy that is captured by the canopy of leaves, consider where the water comes from. The tree's roots draw from the soil the water required for photosynthesis. The ultimate source of that ground water is rainfall (and also snowfall in many parts of the world). What is it that fills the atmosphere with a never-ending input of the water vapor that falls as rain and snow? It is the sun, of course; its radiant energy constantly forces water vapor from the surface of every body of water the world round, including bodies of frozen water such as glaciers and snow fields.

It is a long trip in time from the hearths at Chou-kou-tien to the coal mines and oil fields of today's industrial world. Yet today's fossil fuels are also the products of solar energy. Just how petroleum and its associated natural gas were formed is not yet certainly known, although it is clear that organic matter was present at the starting point. With coal, however, the geologists can tell us not only how many millions of years ago the fossil beds were formed, but can actually identify many of the woody plants that were buried and transformed into fuel. Thus the coal fires that powered the Industrial Revolution and feed our boilers and furnaces today also had their origin in solar energy.

We can go further. The power derived from nuclear fission, a process unknown a few decades ago but as familiar as your power company's nearest atomic generator today, is also a gift of the sun, although a less direct one. The heavy, fissile elements in the atomic plant's fuel rods were created in the stellar furnaces that gave birth to our planet. Only when fusion power is achieved will man have at his disposal a source of energy independent of the sun, and even then he will have achieved his goal by creating miniature suns of his own.

A sequence of solar assistance is shown in this diagram. At the top the rays of the sun are providing the energy that causes plant growth by photosynthesis. At the same time more solar energy is pumping water vapor out of a pond; when the vapor condenses and falls as rain, perhaps in a quick storm (*middle*), it will provide the trees with some of the water they also need for growth. Such a storm will also knock off a few dead limbs and branches that men (*bottom*) can gather and burn for light and warmth; the fire is, of course, solar energy in storage: wood that the sun helped to grow.

Just what is energy? The word is used metaphorically (and carelessly) as often as it is used rigorously. For example you feel energetic on a nice morning and Old Jones is a powerhouse of energy, in spite of his age. To find out what this word really means, look it up in a dictionary. Just ahead of the word itself, if you use the right dictionary, you will find a little-known one: "energumen." Its definition is "someone possessed by a demon," and it is taken from a Greek word *energoumenos,* the past participial form of the verb *energein,* meaning "to work in or on," from *"en"* (in) + *"ergon"* (work). Aha! Thus, energy is the capacity to do work.

As the dictionary will show, the capacity to do work is not too different from being possessed by a demon.

So far as what we call "mechanical" energy (as distinct, say, from "chemical" energy) is concerned, this capacity to do work appears in two forms. One form is latent (usually called "potential" energy in mechanics) and the other form is actual (usually called "kinetic" energy). The first might equally well be called the energy of position and the second the energy of motion. The relation between them is usually explained by the example of a pendulum, the swinging weight that makes antique clocks go ticktock.

You don't see many pendulums nowadays (if your local science museum has a Foucault pendulum on exhibit, it's well worth a trip to see it), but you can easily make one by drilling a hole near one end of a yardstick and taping a weight to the other end. Using a nail that fits loosely through the drilled hole, hang the yardstick on a wall where it will be free to swing through at least a half-circle. You are now ready to observe the relation between latent and actual energy.

When the yardstick hangs motionless, pointing at the floor, it represents an energy level of zero. If you now exercise muscle power to push the weighted end of the yardstick up along the wall until the stick is parallel to the floor, you will have endowed the pendulum with latent energy. The amount of your endowment can be precisely measured in terms of the vertical distance that the weight has been lifted. Regardless of actual weight or distance, we know the endowment is 50 percent of the maximum latent energy the system can receive; if you raised the pendulum through another 90 degrees until it was pointing at the ceiling the endowment would reach 100 percent.

So far, all you have is latent energy — that is, you have endowed the pendulum with a certain capacity to do work because of its position. Now let go of the weighted end. The instant the pendulum begins its downward swing, the energy of position starts being transformed into the energy of motion, that is, from latent to actual. By the time the pendulum reaches the bottom of its swing, all the energy in the system is actual and none of it latent; as the upswing starts, the transformation begins to run the other way. When the upswing slows to a stop all the energy in the system has become potential once more, only to be transformed into actual energy during the returning down-

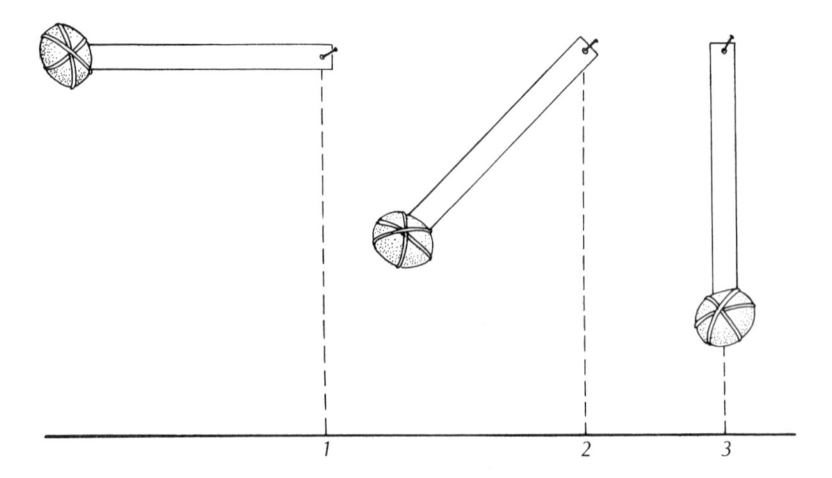

Homemade pendulum: Distinguish between potential and kinetic energy for yourself. Raised weight at the end of a yardstick (*1*) has potential energy stored in it, which is as great as the weight (measured in pounds) times the distance (measured in feet) that it has been raised. When released (*2*), the homemade pendulum swings. The potential energy is transformed into kinetic energy. After several swings the pendulum will come to rest (*3*); it now has neither potential nor kinetic energy and will not have either until lifted again.

swing. And so it goes, back and forth, until the pendulum finally comes to rest as the energy level returns to zero.

Now that the pendulum is motionless, where did your endowment go? Feel the nailhead; it should be warm to the touch. While some of the energy has been spent overcoming air resistance (which is one kind of frictional — or heat — loss) the rest of it has been dissipated in the form of heat where the swinging yardstick has rubbed the nail that supports it. Were it not for these transformations of energy into heat, your pendulum should wag back and forth forever, or at least for as long as the nail and nail hole last.

What you have just seen is important in more ways than one. For one thing, you have seen energy transformed: from latent to actual and back again. This shows that energy can be changed from one form to another. Any number of such changes is possible. For example, when you light a candle or a kerosene lamp you are changing chemical energy into radiant energy in the form of light and heat; when you turn on an electric light, you are changing electrical energy into radiant energy in the same way.

What is less obvious is another fact. While energy can be changed from one form to another with no more effort than swinging a weight or striking a match, it can neither be created nor destroyed. (This truth is known, rather long-windedly, as the Law of the Conservation of Energy.)

Again, your experience with the pendulum has exposed you to both these points. The pendulum ran down; nevertheless the energy in the system was not destroyed. It was changed into another form: heat. And how did the system acquire its energy in the first place? The energy wasn't created; you injected it into the system by lifting the weight at the start. In effect, your chemical energy, expressed as muscle power, was transformed into the latent mechanical energy of the raised pendulum.

And so we have come round a full circle, from the folk at Chou-kou-tien applying muscle power to gather firewood to your own application of muscle power to prepare for the change of mechanical energy from one form to another. Because the capacity to do work cannot be created and cannot be destroyed, what allowed both of you to do this work, ultimately, was that fraction of the sun's enormous endowment of radiant energy that reaches our planet. As we trace man's quest for energy we will encounter the capacity to do work in many forms and see the myriad subtle ways of heat. For the first several hundred thousand years, however, our main concern will be with muscle power and with various ingenious ways to multiply it.

Before we leave the pendulum altogether, however, let me point out that the way this swinging weight transforms potential energy into kinetic energy was a key element in siege warfare almost until the first internal combustion engine — the cannon — came into use in the Dark Ages. The swinging weight, in this case, was a massive timber. It was probably nothing but a roughly trimmed tree trunk at first, but it became fancier and fancier as the centuries passed and eventually its business end was shod with metal. This was a more durable striking surface than bare wood, and someone with a sense of humor and memories of sheep in mating time shaped the metal shoe for his timber in the form of a ram's head, horns foremost for butting. Thereafter this breacher of gates and even of walls was known far and wide as a "battering ram."

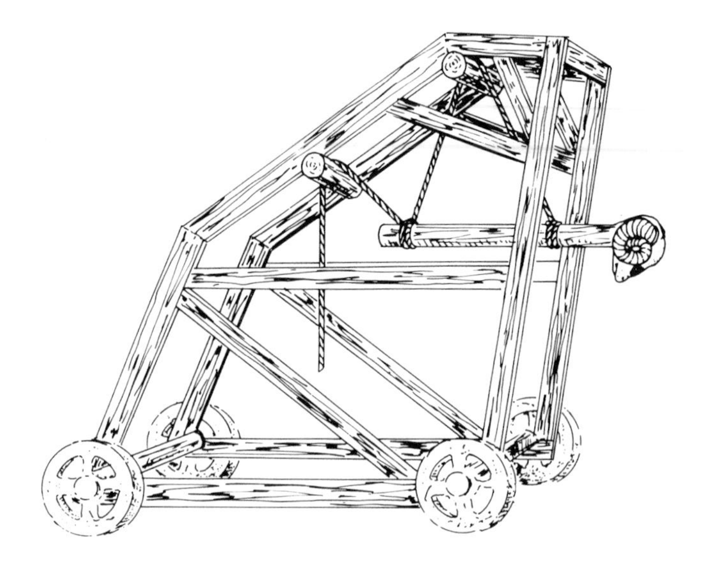

Pendulum principle is what makes battering rams work. This odd-looking one is the author's reconstruction of a kind shown in a famous depiction of a Near Eastern battle scene, Sennacherib's siege of the Judean walled city of Lachish. By pulling on the rope end, the operator gave the log with the metal ram's head potential energy (weight of log times distance raised). By releasing the rope this was transformed into kinetic energy, delivered by the ram's head at the bottom of its swing to any object standing in its way.

Here is a four-Viking-power battering ram that has as much kinetic energy as its crew has endurance.

The pendulum principle? Well, perhaps the first timbers were indeed simply carried by courageous (or careless) runners, as Hagar and his crew may be seen doing here. But when the defenders of the gate can shower the rammers with rocks and arrows and even burning oil, the rammers soon lose heart. The better way is to build a narrow shed with a stout roof and a cover of wet hides for fireproofing. Then roll the shed up to the gate until one narrow end touches the entrance. Your battering ram, of course, is inside the shed, hung on ropes. The rammers heave it back as far as they can (and the ram rises in the same way that the pendulum does, storing up the potential energy injected by the crew's muscle power). At the top of the swing the rammers all let go at once; the descending timber converts the potential energy into kinetic energy as it swings forward, and the entire investment is delivered in the form of a smashing blow of the metal ram's head against the gate. No gate (or wall) holds up for long under such a drubbing! All honor to the Great Pendulum, besieger of cities.

CHAPTER
3

The Hunter and the Hearth

WE LEFT the representatives of *Homo erectus* in a cave in northern China a short while ago, sitting by a fire and eating venison and perhaps even passing the time by hammering nodules of quartz into shape for use as tools. The time was roughly half a million years ago and what makes the scene at Chou-kou-tien different from still earlier campsites known to archaeologists is the presence of fire.

Was it here that men first tamed fire? Almost certainly not. But (and this is an important "but"), until last year, this was one of the first *certain* evidences of men's use of fire. How about meat-eating and toolmaking? By no means are these firsts at Chou-kou-tien. Sites in Africa, more than a million years older than this site in China, have yielded evidence not only for meat-eating but also for toolmaking right on the spot.

You may wonder why, when we are concerned with man's mastery of energy, toolmaking and meat-eating should be particularly important. The short answer is "greater efficiency," but to explain the answer we must take another look at man's biological heritage. The fact is that, up until two or perhaps three million years ago, the ancestors of modern man, like our fellow primates today, were probably vegetarians first and foremost.

We still carry with us the evidence of this plant-eating heritage. Our intestine — the tubing that picks up the work of digestion where the stomach leaves off — is long enough to work well for an eater of raw plant tissue and fiber. Our teeth, too, are broad and flat — more suited to milling and munching than to rending and tearing. Somewhere along the line, however, our ancestors acquired a taste for meat and thereby roughly doubled their efficiency.

C. R. Carpenter, whose studies of the howler monkeys of Panama made him a pioneer in the field of primate behavior, has provided the evidence regarding the inefficiency of an all-plant diet. Carpenter spent months in tropical Asia observing the feeding behavior of the smallest of the great apes, the gibbon. During its waking hours, he found, the gibbon spends half its time eating and the remaining half traveling from its sleeping place to the areas where plant foods are to be found and back again. In effect the ape's entire working day is used up by the hunt for and the consumption of plant foods. The muscles and fat that cover the gibbon's bones represent the conversion of great masses of plant material into nutrient molecules in the ape's stomach and intestine. Like all other vegetarians (technically, the zoologists call them herbivores, which means exactly the same thing), these primates take their nourishment only one step away from the direct use of solar energy.

A meat-eater, in contrast to a vegetarian, lets some other animal do the donkey's work of converting plant tissue into fat and protein. What the carnivore (as the zoologists call it) does is to harvest the end product. When the carnivore at last yawns, stretches, and strolls away from the gnawed bones of its kill, its belly is filled with concentrated nourishment; 1 pound of meat will yield more energy than 20 pounds of bamboo shoots. The carnivore need not kill again for days; eating meat means free time.

It was not necessary for our ancestors to become hunters before they could add meat to their diet. For example, at one of the very early sites in East Africa excavated by Richard Leakey and his associates, the diggers unearthed the scattered remains of a single hippopotamus, together with a number of simple stone tools apparently made in the spot by striking sharp-edged flakes off lumps of volcanic rock. The combination of cutting tools and scattered bones leaves little doubt that some of man's ancestors squatted here, nearly two million years ago, eating hippo, but there is no reason to believe they had first hunted and killed the animal. It is much more likely that they happened upon a recently dead or dying hippo and scavenged it on the spot.

Afterward, bellies full, the meat-eaters may have snoozed in the shade (which is how the great cats use the leisure time they gain by eating meat) or they may have groomed one another, looking for lice for dessert, as the omnivorous baboons do. Or they may have fidgeted. When we look at where man has gone since meat-eating became habitual, the odds favor their fidgeting.

So much for an increase in man's efficiency, measured in terms of spare time man gained by converting himself from an eater of plants to an eater of anything, with a strong bias for red meat. What about the efficiency to be gained by using tools? As Leakey's hippo/artifact site shows, man's ancestors appreciated the advantages offered by sharp-edged flakes of stone when it came to cutting up a large animal. The evidence from this site in East Africa, and from other sites in the same region that are even older, allows us to say that man's ancestors were habitual makers of stone cutting tools as long as two million years ago.

We can say "habitual" without fear of being contradicted because the tools found at the hippo/artifact site are flakes of a kind of volcanic rock that cannot be found anywhere in the vicinity. What is more, one of the larger pieces of volcanic rock used by these ancestral meat-eaters as a source of sharp flakes had been left among the hippo bones, along with the kind of pebble that was used as a hammer to strike the large rock "core" and detach the desired flakes. What all this means is that the meat-eaters actually carried with them both the raw material for making stone tools and the simple primary tool — a hammerstone — needed to make them. Anyone who would so burden himself while tramping about looking for food is a habitual tool-user.

Were stone tools the only tools known to these ancestral

meat-eaters and their descendants? Almost certainly not, although we have to travel hundreds of thousands of years nearer the present before we can produce any evidence on this point. Before we do that, let us briefly exercise our imaginations. One student of early man in Africa, exercising his imagination many years ago, visualized an abundant source of ready-made tools that empty-handed early man could easily have picked up as he went about the daily quest for food. The source was the never-ending supply of clean-picked animal cadavers; the tools would have been leg-bones, horns, and even animal teeth.

Why not? A fibula makes a fine bone dagger and a femur a good club. A long, straight horn could be used as a digging

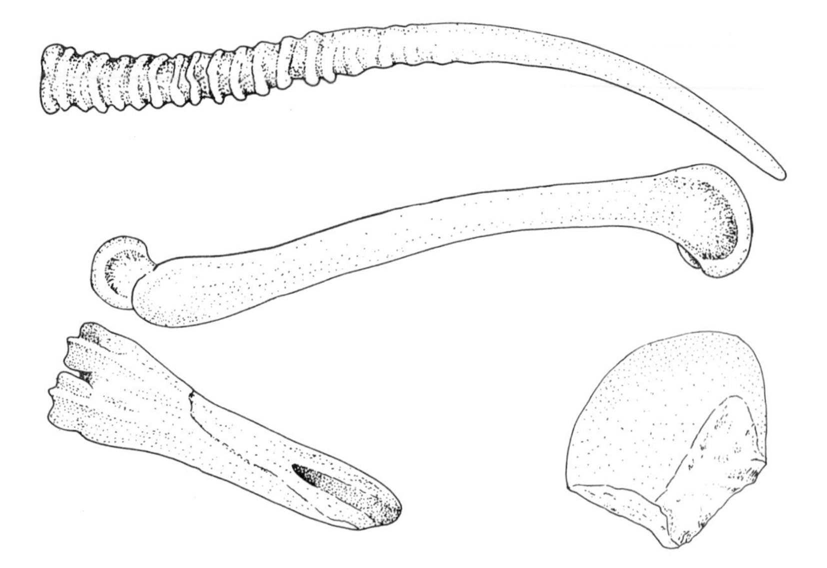

Some of the "tools" that our earliest ancestors may have used, but not made. These are (*bottom left*) a naturally fractured limb bone that could serve as a kind of dagger, a small stone cobble, also naturally fractured, that might serve as a crude knife (*bottom right*), another limb bone usable as a kind of club (*center*), and a horn (*top*) that could serve as a digging stick. Their source, except for the cobble: the animal skeletons scattered across the African landscape.

stick and the canine teeth of a carnivore have cutting edges of a kind. There is no good reason to doubt that man's ancestors used just such tools. At the same time, there has so far been no way of proving that they did so; sites containing the appropriate animal bones, showing the appropriate signs of wear from such use, have simply not been found.

When, in exercising our own imaginations, we think about the kinds of tools man's ancestors might have made out of wood, we face an even emptier past. Really ancient bone tools may someday be discovered because bone is durable and, if buried under the right conditions, can last a long time. (Teeth come close to being indestructible.) Wood, on the contrary, rots rapidly. Only in desert climates, or under highly unusual conditions of burial, is wood likely to be preserved for more than a few hundred years.

This fact need not cripple our imaginations, however; it only means that what we can imagine is nearly impossible to prove. For example, what pieces of wood might you select from the windfalls along the edge of the forest, rather than venturing out on the savanna empty-handed? Something to dig with would be useful. A stout stick with a point at one end would be a great help in grubbing up edible roots or laying open a rodent's tunnel. Something to strike with would be useful too, if only to crack open the shell of a tortoise. The same stout stick could serve both purposes and also, if not too cumbersome, might be thrown with deadly effect at a scampering rabbit.

No good reason exists for doubting that our empty-handed forebears did just what we would do today, selecting multipurpose lengths of deadwood, hardened by months of seasoning on the ground. We can even imagine that the use of such wooden tools (and even animal bones, horns, and teeth) long preceded the use of stone. The time had to come, however, when the ancestors of mankind realized that wood and bone were not enough. We can even imagine what that moment was like.

I will now invent such a scene; it presents ancestral man in the act of discovering his need for a cutting tool. The scene is made out of whole cloth but it can be realistically set in Africa because the earliest evidence of stone tools is found there. Ready? Here we stand on the wooded fringe of the game-rich savanna. There is our hero, a little farther out in the open, crouching on a grassy knoll. He has no eyes for the sun-drenched miles of plain that stretch before him, dotted with grazing animals. This is because the slope just beyond him is a rabbit warren and his attention is fixed on the dark mouth of a warren tunnel. He waits, his wood throwing-stick poised, hoping for a careless bunny to emerge.

But what have we here? Out on the plain there is dust and wild movement! A danger signal: some carnivore is on the prowl. Our hero is suddenly all attention. The rabbits can wait; the big cats will hunt down two-footed game as readily as they hunt antelope or zebra. He watches intently as the spooked herds mill about. Will he have to run for the shelter of the forest fringe? No; the chase is about to end as quickly as it began.

Cleverly driven, a string of antelope moves by on the run, unaware that they are being herded directly toward the big cat's partner in the hunt. The ambusher leaps and strikes; one antelope is down, neck broken. Before the last of the string has passed, the lioness strikes again. This time she is off balance. Her prey stumbles, recovers, and bounces away after its fellows, its torn flank dripping red. The cat does not pursue; her partner has come up from the rear and they join company to devour the kill.

The carnivores' feast is too close to our hero's lookout for comfort. He backs off cautiously, abandoning the rabbit hunt, and heads for his home place, a tree-fringed bend in a stream near the forest edge. Not far along the way he sees that the grass is marked with blood. The wounded antelope has passed this way, lagging behind its fellows. Our hero turns off to follow its track.

A long time goes by and our hero is far from his home place before he finds the dying animal fallen in the grass. As it tries to regain its feet, our hero bludgeons it to death with repeated blows of his throwing-stick. Here is a bonanza of meat, better than a hundred rabbits or twice a hundred turtles! Our hero and the half-dozen fellow hunters who share his home place at night could feed themselves to the point of stupor on this much raw flesh before darkness falls and the carcass must be left to jackal and hyena. If only, that is, our hero could cut it up.

Quite aside from the fact that the carcass is too heavy to carry or even to drag back to the home place, our hero has no way of butchering the antelope into edible, let alone portable, pieces. See him vainly poking at the quarry with his blunt throwing-stick! Rabbits are one thing; a rabbit or two can be carried home whole and then broken up with bare hands and teeth. But what can our hero do with 150 pounds of antelope? Even if his fellow hunters were beside him, there is no way to slice through sinew and muscle. Perhaps one of his companions, who has a bone dagger, could have ripped open the belly and exposed the sweet liver and the fat-encased kidneys tucked away beneath the smoking tripes. Even the dagger-carrier, however, could not cut up the carcass.

Alas! Our hero can only use his thumbs to gouge out the tasty layer of fat that lies beneath the antelope's eyeballs and, somewhat heartened by this small snack, head back for the distant home place before dark overtakes him. What does he have to show for a day's hunt? Nothing. A dim light begins to flicker in our hero's mind. Why not invent the knife?

All of this scene, of course, is purely imaginary. For example, the rabbit-hunter may not have shared a home place with other hunters. He may not even have had a home place. There can be no doubt, however, that at some point in their unrecorded history our remote ancestors recognized their need for a sharp-edged tool that would not only cut meat but also slice through hide and sinew, plant stems, wood and other materials. We have already seen, at Leakey's hippo/artifact site, that the response to this need had become habitual in East Africa some two million years ago. Do not ask how long before then the first wood or bone bludgeons, digging-sticks and throwing-sticks came into use. It would be almost miraculous if any evidence of these "firsts" ever comes to light.

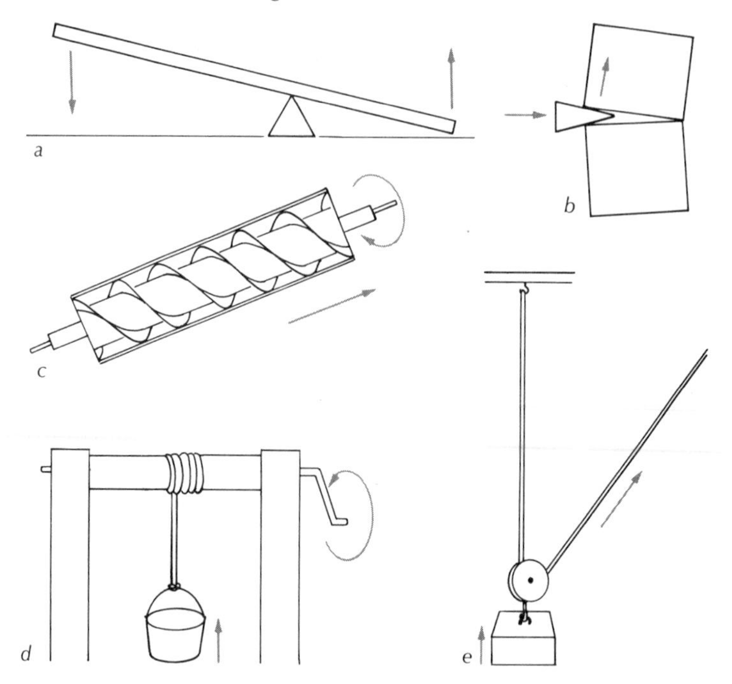

The Greeks' five "simple" machines. The first (*a, top left*) is the lever. Of its great power, Archimedes is said to have remarked that, given a long enough lever and a suitable fulcrum (the shaded triangle shown here), he would undertake to lift the earth. Man's first lever probably was a digging stick. The second (*b*) is the wedge, shown here splitting a block of wood. Man's first efficient wedges were stone chisels, axes, and knives. The third (*c*) is the endless screw; the version seen here is Archimedes' invention, a water-raising machine. The fourth (*d*) is the wheel and axle (for example, the potter's wheel). Shown here is the version we call a windlass; its "wheel" is the horizontal barrel between the two uprights; the axle on which it revolves is at the end opposite the crank. The fifth (*e*) is the pulley. The latter three "simple" machines are inventions of man's recent past, the potter's wheel probably being the oldest.

In terms of energy, what did our earliest forebears gain by using tools? For the most part, more of what they had already gained by adding meat to a vegetarian diet: spare time. It is less work to winkle an edible root out of the soil with a digging-stick than it is to grub it out with bare hands. Doesn't this tool seem familiar? It should; it is nothing other than one of the Greeks' five "simple" machines — the lever. At a guess, using such a mechanical aid, the productivity ratio might be four or five stick-extracted roots to every hand-dug one. If so, the saving in unexpended time and energy is 75 to 80 percent. If we go on to contrast rabbit-hunting by the bare-handed rundown-and-snatch method with rabbit-knocking by the throwing-stick method, the saving might approach 99 percent. It is not easy to calculate differences of this kind, but obviously any increase in efficiency means more of a saving in the time and energy needed to get something done, and thus more time and energy available for other concerns.

Actually, the individual saving in time and energy gained by the first use of tools instead of bare hands, was probably less important than the really enormous increase in efficiency brought about by the changes in human social behavior induced by tool use. For example, in my fictional scene of the hunter who had no knife I casually assumed not only that he ended his daily food quest by returning to a home place but also that he would meet half a dozen fellow hunters when he arrived there. I could also have assumed that most of these fellow hunters would be accompanied by mates, and that the home-place group would include a number of infants and immature offspring of these matings. In my assumption, the "why" of such an aggregation was not even examined; it was as much a product of the imagination as the lions' pursuit of the antelopes.

Once our ancestors held cutting tools in their hands, however, the why of such social groupings becomes obvious. Two can fare far better than one and a dozen far better than two. When butchering became possible, sharing became practical.

And when eating becomes a group practice it is no very great step further to begin group hunting too.

Is this picture of a home place where ancestral hunters gather to share their quarry purely imaginary? It would have been only a few years ago. Today, just such a site has been found and excavated. The hunters had selected a pleasant place, a sandy stream bed, shaded by trees. The stream bed may have been dry most of the year, but water was available from a nearby lake. No direct evidence of how many hunters used the home place was found, but among the more than 400 bits of animal bone uncovered by the excavators were the identifiable remains of a giraffe and a hartebeest, as well as such smaller game as porcupine.

If the occupiers of this home place actually hunted down animals as large and as dangerous as a giraffe (giraffes look quaint and lovable in the zoo but they can kick a lion to death), rather than butchering a dead one, then the odds greatly favor their being group hunters, and the practice of group hunting would suggest that at least four adults were present. (Note that I do not say "men." At this stage in prehistory, with no home fires to keep burning, it seems likely that both males and females hunted, and not just the adults but the older sub-adults as well.)

Actually, we can't be sure that these food-sharers were also group hunters. The site, another of those discovered in Kenya and excavated by Richard Leakey and his associates, is not far away from the hippo/artifact site and is about as old. The toolmakers who sat down to eat the hippo were almost certainly scavengers rather than hunters, and these neighboring giraffe-eaters may well have been the same. Thus, both as to the question of how many hunters used this home place and as to the question of their hunting strategy, the archaeological evidence is not clear. Still, in terms of man's increasing efficiency, the site teaches us quite a lot. At the very least, these remote ancestors regularly came to the same place to share the food that their possession of cutting tools allowed them to butcher and carry home. Foreshadowed here is the nucleus of human society and,

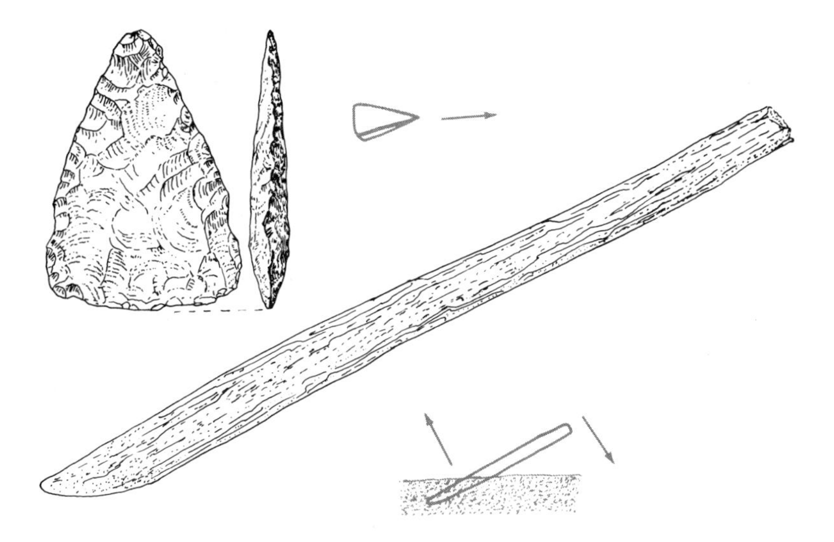

Two of the five "simple" machines, as utilized by Ice Age man only a few hundreds of thousands years ago. At top is the equivalent of a wedge, an Acheulean hand ax from France that was probably used more to "wedge" apart meat from bones than for splitting wood. Below it is the equivalent of a lever, a stout length of digging stick used to pry up edible plants. No digging sticks of Ice Age date have yet been found; this is a 19th-century one. The discovery in England of a "spear point" made of wood at about the same time as the French hand ax was made, however, proves that Ice Age men were users of wood artifacts; why not digging sticks? What is more, a 1981 study has now suggested that *H. erectus* or even pre-*erectus* inhabitants of Kenya some 1.5 million years ago were working on wood artifacts, although no artifacts survive.

with it, the important economies of effort that come with group living. Energy that is saved can be invested elsewhere.

The flakes of volcanic rock found at Leakey's Kenya sites, together with similar simple stone tools unearthed in other parts of East Africa, are among the earliest known examples of our ancestors' handiwork. Today we have little difficulty in recognizing them as purposely manufactured artifacts and not merely accidental debris. Less than 150 years ago, however, no one would have given them a second glance. At that time, not even the flint tools of the Old Stone Age — as far superior in workmanship to the Kenya flakes as the space shuttle is to the Wright brothers' biplane — were believed to be the works of man. Scholars called them ceraunias (after *keraunos,* the Greek word for thunderbolt) because they were supposed to have formed spontaneously when lightning struck the earth. Some experts even wasted energy speculating about the conditions of soil moisture and atmospheric humidity that were required for the lightning stroke to produce the stones. Others, confident in the common wisdom that lightning never strikes twice in the same place, carried the thunderbolts home to set up around the house as a kind of insurance against storm damage.

Some decades before the American colonies won their independence, an English antiquarian published a drawing of a large stone chopper that had been found along with the bones of an elephant in a London gravel pit. The tool, he said, was "a flint lance like unto the head of a spear," and had every appearance of great age, for surely elephants had not roamed England in man's memory. The scholars quickly settled this poor fellow's hash. Had not the Emperor Claudius brought elephants with him when he Romanized Britain? And could not some of the native Britons, too poor, perhaps, to own bronze or iron weapons, have come to battle the invading Legions armed with nothing better than stone-tipped spears? Surely the gravel-pit discovery was nothing more than a chance relic of the Roman conquest.

More than a century later, a French civil servant at last forced the skeptics to recognize stone tools as the handiwork of early man. "Antediluvian" man was what scholars of the 19th-century called our ancestors. This is to say "pre-Flood," the flood in question being that deluge survived only by Noah and his shipmates. While all this seems rather charmingly old-fashioned to us, even many educated folk in the mid-19th century believed without doubt that a wrathful God had once drowned all of mankind except for that chosen handful. Nor did it end there, the pious of that era also accepted on faith the estimate that the world had been created in the year 4004 B.C., for this was the date that had been calculated two centuries earlier by a renowned Biblical scholar, James Usher, Archbishop of Armagh. But back to the stone tools.

The French civil servant who challenged dogma carried an imposing name: Boucher de Crèvecoeur de Perthes. By profession a customs officer like his father before him, de Perthes was an amateur botanist and dedicated naturalist. In 1825 he was appointed to succeed his father as Collector of Customs at Abbeville, the Channel port at the mouth of the Somme River. It was some five years later that he first found in the gravel quarries of the river valley rough flint artifacts that he recognized as the work of man. De Perthes kept his eyes open thereafter but more than 15 years were to pass before he found, at Menchecourt, a gravel deposit that included a flint tool in association with the bones of elephant and rhinoceros. He published the first of a series of volumes the next year, 1847, under the title *Celtic and Antediluvian Antiquities,* but his claims were generally ignored. At least no one suggested that the bones were left over from some Roman circus; by this time scholars knew how to tell ancient animal bones from recent ones and the great French paleontologist Cuvier vouched for the antiquity of the Menchecourt remains. Nonetheless, another decade passed before de Perthes won recognition.

Oddly enough, it was not the French who first realized what de Perthes had proved; his only champion at home was a physician and amateur antiquarian in neighboring Amiens. Instead, his support came from a Scot and an Englishman, both with considerable experience in geology. The Scot was Hugh Falconer who, after taking a degree in medicine at Edinburgh, had gone out to India. The Englishman was Joseph Prestwich, who first encountered geology at the University of London and pursued the subject for the rest of his life while making his day-to-day living as a wine merchant.

While in India, Falconer had explored the rich fossil beds in the Siwalik Hills northwest of Delhi. His work there and elsewhere in India made him familiar with the fossil remains of, among others, mastodon, elephant, and rhinoceros. The two men met and became friends after ill health finally forced the Scot's return from India to London. In 1858, Falconer visited de Perthes at Abbeville and examined his finds. At Falconer's insistence Prestwich made the trip to Abbeville the next year; both men agreed with de Perthes that the stone tools he had found in association with the bones of extinct animals were indeed the work of men.

Prestwich thereupon prepared a report, noting that the association was proof that man had existed at least as long ago as the great mammals of the Ice Age — that is, long before 4004 B.C. His paper appeared in *Philosophical Transactions* in 1861. By 1863 the French Government had awarded de Perthes the rosette of the Legion of Honor and his work had achieved international recognition.

But if men had made these stone tools, where were the human bones? By now de Perthes was known to every gravel quarryman in the Somme Valley. When it became public knowledge that he would pay a reward to anyone who happened upon ancient human remains, it can hardly come as a surprise that just such a discovery was soon made. The gravel pit that was its scene was even conveniently near Abbeville, at Quignon Mill. There de Perthes was led to a fresh excavation that had exposed, in addition to some flint tools, a human lower jaw. De Perthes, who was no expert on human fossils, was happily prepared to believe the jawbone was as ancient as the tools.

This, too, should hardly come as a surprise. Up to this time very few human fossils had been discovered. Moreover, scholars challenged the antiquity of the few bones that had been found. For example, in 1832 a Belgian antiquarian, P. C. Schmerling, who was fossicking in a cave on the banks of the Meuse River, not far from Liège, had found a human skull and other human bones mixed in with the bones of mammoth, rhinoceros, cave bear, and hyena. The cave deposit had been disturbed, however, and skeptics suggested that the human remains were much later in age than the animal bones. A quarter-century later, a human skull cap, two leg bones, and two arm bones had come to light in the valley ("thal" is the German word for valley) of Neander, near Düsseldorf; amateurs had made the discovery in a cave, Feldhofen Hole. Although, today, Neanderthal man is virtually a household word (to call someone a Neanderthal is to accuse him of ultra-right-wing views), most of the 19th-century anatomists who examined the newfound skull cap suggested it was that of a modern man who had suffered some bone-deforming disease either in infancy or later in life.

The experts were wrong, of course. We will hear more about the Neanderthals later. Unfortunately for the progress of learning, de Perthes, too, was wrong. The Quignon jaw was soon proved to be modern; it had been "planted" among the flints by someone who wanted de Perthes's reward. The fraud did much to cloud the whole issue of man's true antiquity. Two decades more were to pass before another discovery in Belgium gave solid evidence that humans and Ice Age animals had lived at the same time.

In 1886, digging in still another cave, Betche aux Roches, near Namur in Belgium, Maximin Lohest and Marcel de Puydt found two nearly intact human skeletons buried 16 feet below the floor of the cave. Associated with the human remains were flint tools of a distinctive type that in the years ahead were often to be unearthed at sites where Neanderthal fossils are found. The bones of Ice Age animals were also present: mammoth, woolly rhinoceros, and cave bear.

Today the two individuals, called the Spy (pronounced "spee") fossils after the name of the commune where Betche aux Roches is located, are recognized as Neanderthals in good standing. When they were first found, however, they were classified by the experts as the lowest form of humanity yet known, "near to the apes," and so forth. Still, the Spy pair silenced the skeptics. There they were, the two of them alike as peas and 16 feet down amid Ice Age fossil animals and flint tools. At that depth they could scarcely have been modern burials, and even the most hardened doubter was reluctant to claim that both had by coincidence suffered the same deforming disease. Here was Ice Age Man!

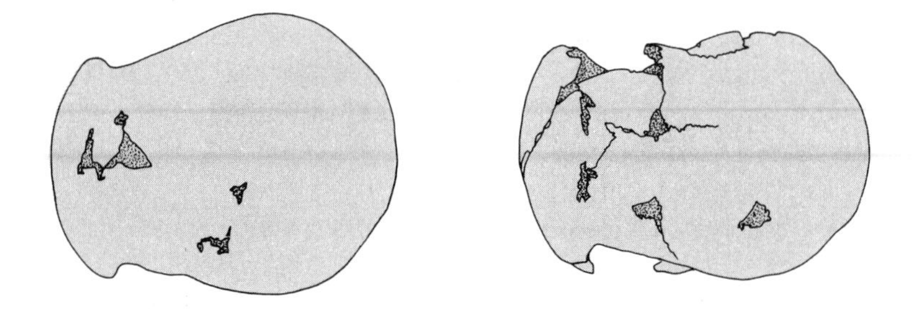

The two Neanderthal skulls found in a cave (Betche aux Roches) near Namur in Belgium in 1886 as illustrated in the discoverers' report. Like the original skull from the cave in the Neander valley (Feldhofen Hole) near Düsseldorf in Germany, these did not look like modern skulls. For one thing their brow ridges (*left*) were too large and prominent. Experts had dismissed the Neander skull as modern but diseased. The odds against two modern men, suffering the same supposed disease, being buried together, 16 feet deep, along with the bones of Ice Age animals, were so extreme that the pair, known as the Spy fossils, were accepted as genuinely old.

The Spy discovery actually rehabilitated the fossil from the Neander valley cave and incidentally caused an antiquarian at Gibraltar to re-examine a skull found at Forbes quarry there in 1848. The Gibraltar skull was thus belatedly recognized as both ancient and Neanderthal.

Incidentally, for those who like romantic endings, de Perthes's supporter in the 1860s, Falconer, had died only a few years later, exhausted by the diseases he had contracted in India. His friend Prestwich, then a bachelor in his 40s, wooed and won Falconer's niece, Grace Milne. Prestwich eventually gave up the wine business to devote his full time to geology. He was honored by appointment to the chair of geology at Oxford and up to age 80 and beyond could be seen puttering about the hills of Kent, the county to which he retired, collecting tools of the Old Stone Age. In 1895 he was knighted, an act that gave Falconer's niece the right to call herself Lady Grace. Once knighted, Prestwich promptly died, rich in years and honors. But I digress.

We were concerned, before this little detour into the discovery of man's true antiquity, with the increase in efficiency and the energy saving that accompanied our ancestors' shift to a diet that included meat. Out of the earliest human behavior we can reconstruct — those patterns of social living begun nearly two million years ago — must have arisen the practice of group hunting, a development that added another welcome bit of efficiency to our forebears' actions.

It is not known how long ago the practice of group hunting began. The first tentative evidence of the practice does not appear until middle Old Stone Age times, perhaps half a million years ago and perhaps less. The evidence consists mainly of the bones of animals that have clearly met violent ends. From the kinds of animals represented — large, medium, or small — and from their numbers — a single beast, a few or many — one can imaginatively reconstruct various possible hunting strategies and then search for additional evidence indicating which one of the possible strategies may have been used. This we will do shortly, at the same time we take note of how man's toolkit is being enlarged. But first we must examine something else.

What single factor provides the major contrast between the East African streamside picnic ground of some two million years ago and the North China cave camp that was occupied perhaps 1.5 million years later? Is it the quality of the stone tools? No. The tools from Africa were hammered out of volcanic rock and the tools from China were made from quartzite pebbles, but this is only a difference in raw material. Their users could whittle, slice, and crush things about as well with either kind of stone. The tools from China are also somewhat better made, but this might reflect nothing more than the fact that quartz fractures more neatly than basalt.

Is it the variety of menu? No. Although hard evidence for the consumption of plant foods has yet to be found in East Africa, no one can possibly imagine that these early ancestors of man ate nothing but meat. At the same time, the different kinds of animals they carried to their picnic ground far outnumber the different kinds carried home to the cave in China, where evidently the chief game animal was deer.

You have guessed the difference. When the sun set at the streamside grove in Kenya the picnic ground grew dark whereas at Chou-kou-tien a flicker of firelight held back the night. *Homo erectus* could keep a fire alive but the hominids of East Africa, so far as the evidence shows, could not. One can imagine that the Africans' only experience with fire was what the game they fed on also experienced: the spontaneous wildfire of the savanna grasslands, kindled irregularly by electric storms.

Man's myths tell many stories about "culture heroes" (as they are called by scholars): figures larger than life and often semi-divine, who taught the earth's lesser folk how to plow or make wine or play the pipes. The greatest of all these heroes among the ancient Greeks was the Titan Prometheus, who brought mankind the gift of fire.

In one version of the myth, a fellow Titan — Zeus, ruler of the heavens — ordered Prometheus and one of his brothers, Epimetheus, to populate the then still empty earth. Epimetheus, a goodhearted but rather stupid Titan, set about at once to create animals. His supply of creative raw material was limited but he rationed it out carefully. To some of his animals he gave wings for freedom, to others, strong talons and sharp teeth for hunting, and to still others, hard shells for shelter or swift limbs for flight.

Prometheus caught up with his brother just as Epimetheus used the last of the creative stuff by giving courage to the lion. Nothing daunted, the wiser brother gathered a handful of clay and made man and woman in the image of the gods. He gave these first humans an upright posture so as to distinguish them from his brother's creatures. In this way, he pointed out, while the animals' heads were all turned earthward, man would face the stars. And he did one thing more. Selecting a dry tree branch, Prometheus carried it up to the chariot of the flaming sun, kindled it there, and came back to man bearing the gift of fire. With fire in his possession, Prometheus said, man would win domination over the earth and all the riches that it contained.

The Greeks who embroidered this moral tale were themselves living in the plenty and contentment of the Bronze Age (if not later), when fire had been a commonplace human possession for at least half a million years. I find it fascinating that they, so far removed from the countless millennia of cold and darkness experienced by our fireless ancestors, could have recognized the importance of fire so clearly and have marked it down as the divine gift that would let man conquer nature. For it is by using fire — the light-giver, the warmth-provider — that man first seized for himself a unique part of the sun's abundant energy.

To repeat briefly, all plant and animal life on the earth draws its energy directly or indirectly from the sun. When we eat an apple we are stealing a small packet of sunlight (and, incidentally, "burning" it in our stomachs). When we build a fire of apple boughs we are stealing a considerably bigger packet of energy derived from the sun. This is equally true of materials other than wood — plant oils and animal fats for the lamps and candles of yesterday, or the fossil fuels, petroleum and coal, that produce our steam and render our lamp filaments incandescent today.

With the first evidence of man's using fire, at Chou-koutien, we have reached a point in our journey where a more exact timing of events is possible than the rough jumps of a million or a half-million years we have used up to now. To say "more exact" does not, however, mean anything like exactly exact. The problem is that the slice of time at the end of the Cenozoic Era, the Pleistocene epoch, the most recent epoch of that Era, was once thought to have lasted fewer than one million years. Those years were conveniently packaged into subdivisions according to the successive advances and retreats of the great ice sheets that covered much of Europe during the Pleistocene Ice Age. Each advance and retreat altered the landscape in ways geologists can recognize. Indeed, the names given to the last four oscillations come from locations in the Alps where the evidence of past ice action is still apparent: Günz, Mindel, Riss, and Würm. (A still earlier glacial episode, the Donau, does not concern us here.)

These four oscillations and the interglacial periods between them can be recognized in ways other than how the landscape was altered. For example, the animals and plants that thrived in any particular part of Europe during a time of maximum cold were completely different from those found when the climate was warm. Animal fossils and the preserved pollen of shrubs, grasses, and trees therefore indicate the state of the climate at the time they lived.

Today it appears that the Pleistocene epoch lasted at least twice as long as people believed only 50 years ago: perhaps more than two million years rather than only one million. Do the new findings stretch out the time between Günz and Würm? Not very much, actually. The greater part of the expanded time is assigned to a pre-glacial period called the Villefranchian. Starting with the Recent (10,000 B.C. and later) and working backward, the timetable is generally as follows: 12,000 years ago; the end of Würm III, the third and final Würm ice-sheet advance. Between 75,000 and 125,000 years ago: the long interglacial period between Riss and Würm. Between 125,000 and 300,000 years ago: Riss II and Riss I. Between 300,000 and 400,000 years ago: the short interglacial period between Mindel and Riss. Between 400,000 and 600,000 years ago: Mindel (the hunters of Chou-kou-tien would thus have flourished in about mid-Mindel times). Between

YEARS BEFORE PRESENT	glacial sequence (EUROPE)	GEOLOGICAL (archaeological) periods	humans
3,000,000 to 1,900,000	DONAU Donau-Gunz interglacial	BASAL PLEISTOCENE; VILLEFRANCHIAN (first chopper and pebble tools)	(E.R. 1470 and other possible precursors of *Homo* in East Africa)
to	GUNZ I GUNZ II	BASAL PLEISTOCENE; (Early Lower Paleolithic; Chellean tools and earlier bifacial tools; pebble and chopper tools)	*H. erectus* in East Africa
500,000	Gunz-Mindel interglacial		*H. erectus* in China
	MINDEL I	(early Acheulean tools; pebble and chopper tools)	*H. erectus* in Java
to	MINDEL II		*H. erectus* in China (Vertesszollas)
	MINDEL III	LOWER PLEISTOCENE (Lower Paleolithic: middle Acheulean tools; pebble and chopper tools)	(Swanscombe)
275,000	Mindel-Riss interglacial		(Steinheim)
	RISS I		(Fontechevade)
to	RISS II		
	RISS III	MIDDLE PLEISTOCENE (Late Lower Paleolithic: late Acheulean tools)	*H. sapiens neanderthalensis* (by 100,000?)
100,000	Riss-Wurm interglacial		
	WURM I		
	WURM II		
to	WURM III	(Middle and Upper Paleolithic: advanced flint and other tools; Mousterian tools)	*H. sapiens neanderthalensis*
	WURM IV	UPPER PLEISTOCENE	*H. sapiens sapiens*
12,000 to 8,000	(post-glacial)	HOLOCENE (Mesolithic, Neolithic)	*H. sapiens sapiens*

600,000 and 700,000 years ago: another short interglacial period separating Mindel and Günz. Allot another 200-odd millennia to Günz and you've accounted for the better part of a million years.

All these dates are only estimates. They were different in the past and they will surely be revised in the future. Any change, moreover, will almost certainly be in the direction of making everything older. We need not fret about this. What we have in this timetable — running successively from the Villefranchian, some two million years ago, to Würm's end, a mere 12,000 years ago — is a firmly established *relative* chronology. The actual date of Riss II may be changed tomorrow, but Riss II will always come after Riss I.

Hereafter, where appropriate, you will find references to this geological chronology. You will also encounter a parallel relative chronology, not based on the oscillations of the Ice Age but marking instead the successive achievements of our ancestors. This cultural chronology begins with the earliest phase of the Old Stone Age: the Lower Paleolithic. It continues through the Middle and Upper Paleolithic and a subsequent transition period, the Mid-Stone Age, or Mesolithic, which appears in Europe after Würm III. Next is the New Stone Age, or Neolithic. Thereafter, following another transition period that is evident only in some areas of the Old World (the Copper-Stone Age, or Chalcolithic), comes the Bronze Age and the beginning of recorded history, when dates start to be dates, at least in some places. So much for better timetables.

What has prehistory to tell us about man's use of fire in the time since *H. erectus* bands warmed their shanks at Choukou-tien? We can start on a fine autumn day in late September, 1911, in the company of an English amateur geologist, Samuel Warren, who is engaged in his favorite occupation: searching out fossil bones and other natural curiosities along the Essex seashore.

Warren, together with other members of the Essex Field Club, had gone to a North Sea resort town, Clacton-on-Sea, that day to inspect a rocky outcrop of freshwater clays that had been exposed by wave action to the west of Clacton Pier. The most prominent fossils at Clacton were the bones of extinct elephants,

(Associated artifact)

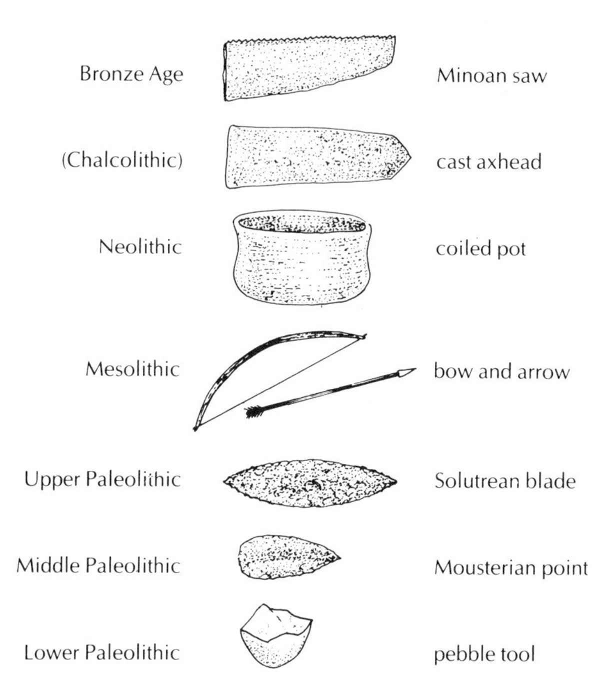

Major divisions and subdivisions of prehistory are characterized here by drawings of key artifacts of the period. Thus a simple stone pebble tool is representative of the most ancient subdivision of the Old Stone Age, a Mousterian point is representative of the next subdivision, and a Solutrean blade of the youngest subdivision. A bow and arrow mark the Middle Stone Age and a coiled pot the New Stone Age. The Copper/Stone period, not found in many parts of the world, is represented by a cast copper axhead and the most recent major division of prehistory, the Bronze Age, is represented by a saw blade. The art of writing was discovered early in the Bronze Age in the Near East, thereby closing the era of prehistoric times.

but on this occasion Warren spied what appeared to be the tip of a straight, pointed stick that was protruding from a peaty seam in the clays. Warren dug the stick loose and found himself holding a man-made object: the tip of a broken yew shaft, 15¼ inches long, tapered to a point by scraping, and less than 1½ inches in diameter at its thickest point.

Warren had discovered the oldest wooden artifact yet unearthed anywhere in the world. Needless to say, he returned to Clacton-on-Sea again and again over the next few years, and by the time he and his colleagues published an account of their various discoveries in 1923 it was clear that the flint tools they had also found among the fossils at Clacton represented an Old Stone Age "industry" (that is, a consistent style of toolmaking) quite distinct from the one de Perthes had discovered. (His had been named Abbevillian, after the town in which he lived.) The flint industry of the Essex shore, which is about 250,000 years old, was given the name Clactonian, and so Warren's 15-odd inches of Paleolithic wood came to be known as the Clacton spearpoint.

When the yew shaft was first described, many people were still familiar with the technique of fire-hardening, a means of strengthening a wood point by charring it in a fire and then scraping away the charred layer. The scorched wood underneath the charred layer is harder than the unheated parts of the shaft, probably because the natural moisture has been baked out of it. In any event, it was reported that the Clacton spearpoint had been fire-hardened. If this were true, Warren's find would have represented two "firsts": the oldest known wood artifact, and the earliest certain evidence of man's use of fire. (The hearths at Chou-kou-tien had not yet been discovered.) In 1952, however, Warren decided to give his celebrated spearpoint to the British Museum (Natural History) and the artifact was re-examined with great care. The results were finally published in December 1977: the examiners could find absolutely no evidence that the yew pole had been fire-hardened.

So there we are: another myth deflated. The first certainly

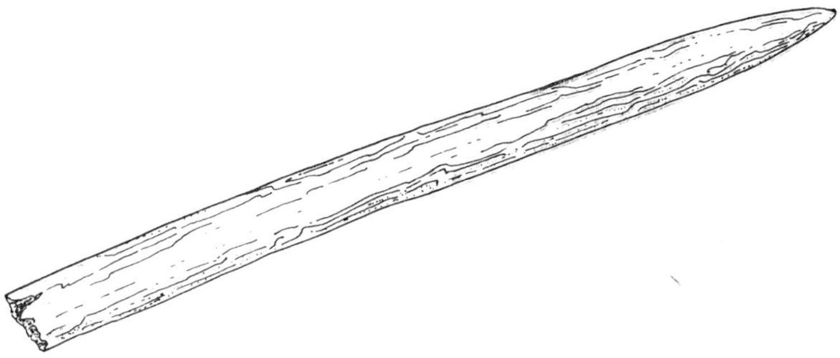

Fire-hardened or not? This is the 15¼-inch broken yew shaft found by Samuel Warren at Clacton-on-Sea in 1911. It is less than 1½ inches in diameter at its thickest and was tapered to a point by scraping. The tools associated with it are estimated to be about 250,000 years old, which makes the shaft the oldest surviving wood tool. It was thought to have been fire-hardened and therefore possibly a spear. Re-examination a few years ago, however, could discover no trace of fire-hardening and so the Clacton "spear" could have been nothing more bloodthirsty than a digging stick.

fire-hardened spearpoint is not from Clacton after all. Instead, it is one that was found not long ago at Lehringen, in Lower Saxony, in association with elephant bones and flint tools typical of the Riss-Würm interglacial period. The Lehringen point is thus probably less than 100,000 years old. In terms of man's prehistory, this is almost yesterday. The Clacton failure also leaves an unpleasantly large gap of perhaps half a million years between the first (in China) and the second (in Europe) certain evidence of man's use of fire. Is there no other evidence, even indirect, that fire was man's helper, if not his servant, during all those hundreds of intervening millennia?

Just one. Another English Old Stone Age site, Hoxne in Sussex, contains a stone-tool industry about as old as the Clactonian industry but of quite a different character. The industry is named Acheulean after a French site, St. Acheul, where its distinctive tools were first recognized. The Hoxne finds are associated with ancient lake beds that have long been dry; the lake-bed sediments contain pollen from Ice Age plants that grew along the shore. The excavators at Hoxne have analyzed these pollens. They were found to be typical of the oak forest that was characteristic of the mild European environment during the Mindel-Riss interglacial period.

That is, the pollens were typical except at one point. In certain midway strata the forest pollens thin out and in their place are found the grass pollens typical of open ground, together with traces of charcoal. Why? One answer might be a shift from forest to steppe — that is, from temperate to cooler climate. Such a shift could bring a change in vegetation of this kind. But the climate was not so changeable during the Mindel-Riss interglacial. The traces of charcoal suggest another answer: forest fires of natural origin. These would have removed the mixed oak cover and grasses could have invaded the burned ground.

Is still a third explanation possible? One such alternative is suggested by the fact that the Acheulean tools are found in the same strata that include the grass pollens and charcoal. If burning was the cause of the forest clearance, could the fires have been man-made rather than natural? The excavators of the Hoxne site believe this to be the case, although they have no evidence to suggest whether the fires were set for a purpose or only by accident. Considering the relatively sophisticated lives that the Hoxne toolmakers lived (we will come to this subject later), I favor the possibility that the fires were set on purpose and, furthermore, that they were set in order to drive game into a hunters' ambush.

So far, then, we have direct evidence of two of the ways early man used fire: the hearths in China and the Lehringen spear. We also have indirect evidence — the Hoxne charcoal and grass pollens — of a possible third use: game-driving. All three uses are certainly important and a probable fourth, cooking, would have made our ancestors' meals much more palatable, at least according to today's standards. None, however, is anywhere nearly as important as night illumination. Our ancestors have had this use of fire at their disposal from the time of the Chou-kou-tien hearths onward, at the very least. Indeed, it now appears that it was at their disposal very much earlier than that. What did this release from the bondage of night mean to mankind?

Consider, if you will, a fair summer night. Once the last hint of crimson has faded from the higher wisps of stratus cloud there is little to do, even under the light of a full moon, except perhaps to go courting or, perhaps more sensibly, to go to sleep. Now, the moon is full, or nearly full, only 5 of the 28 days in the lunar cycle and days are longer than nights only half the year, so that on the average 10 or 11 hours out of every 24 are lost to darkness.

For our ancestors the very first fire changed all this. Within the circle of firelight the flint-knapper could continue to strike flakes from his stone core and shape the core or trim the flakes into the desired form with delicate retouch strokes. The woodworker, who was perhaps the same craftsman as the flint-knapper, could char the tip of a spear-shaft-to-be over the glowing coals or use the firelight to gain a clearer view as he scraped away shavings of wood from a poorly balanced throwing-stick. The men could gather in the circle of light to sketch out plans for the morning hunt; the women, young and old, could ravel bark for cordage, dress skins, or cut fine thongs from a prepared hide. The children — if they were not already, in these remote times, sent to bed early — could creep close to the fire and watch, listen, and learn.

All these actions are simply daylight pursuits, continuing after the fall of darkness. What has the firelight that makes possible their continuation at night added to the efficiency of man? It has allowed a 15 percent stretch-out of the working day; say as little as 5 percent during the short nights of summer, but as much as 30 percent in the dead of winter. If we take 24 hours to represent 100 percent, then man without fire could invest his energies only 50 percent of the time on the average: he was 50-percent efficient. *Fiat pyrolux!* Now he is 65-percent efficient.

This stretch-out of the working day was surely the first and, at least in this remote period of prehistory, the most important result of our ancestors' capture of solar energy. It also seems likely that possession of fire was the push early man needed to set his mind to work on the question of shelter. This is more than possible; it is highly probable, for at this time the means of kindling a new fire were unknown and the loss of a fire could mean months of misery.

A fire in the open will burn during a rain if the supply of dry wood is ample enough, but three stormy days in a row will wet down the woodpile and drown out the hearth. Thus it is scarcely surprising that where the landscape includes rock overhangs, steep bluffs, and, best of all, caves and caverns, the sheltered soil underfoot often contains evidence that men once took refuge there. As we will soon see, it is not impossible to live and hunt in the same area the whole year round, but, in the main, hunters move with the seasons. A fire in the open is fine in summer months, but I do not think it too much to suppose that what kept the home fires burning in wintertime was some kind of natural shelter.

Before men knew how to start a fire, how did they manage to find one? Speculation about this is at least as old as the Prometheus myth. Vitruvius, engineer to the Caesars (we will meet him again), favored the notion that man first stole fire from an erupting volcano. This is perhaps because Vitruvius had seen both Etna and Vesuvius, or perhaps because he was influenced by the Prometheus myth. Volcanoes are not very common, however, nor are they always erupting. A scholar of the 19th century visualized another source of fire, set in a forest scene. As the breezes passed through the treetops, the dead limbs of two adjacent trees chanced to rub back and forth until the heat of friction made them burst into flame. Presumably this was just in time for a passing man to observe the blaze and exclaim, "Just what I've been looking for!"

Nobody can say that either speculation is wrong. Common sense, however, suggests a more plausible alternative: wildfire. On the great veldts of Africa it is a rare day when the horizon is not darkened in one direction or another by the smoke of a wildfire, set ablaze by lightning. Today in the United States the Federal Bureau of Land Management monitors lightning, particularly in the West, where more than one quarter of all fires on public lands are caused by lightning. Thanks to this, some lightning statistics exist. In 1976, the Bureau set up six lightning-detection units in Alaska. They successfully plotted the individual strokes of lightning responsible for 227 forest fires that year. Alaska is very large, to be sure, but this is still an average of more than one forest fire every other day of the year. Worldwide, it has been estimated, lightning strikes at the rate of 6,000 times per minute.

I do not suggest that, if you want to have a wienie roast, you throw away your matches and wait for a wildfire to pass by. There is, however, no reason to believe that lightning was striking the earth any less frequently a million years ago, whereas it is certain that the forests then had not been largely axed away by man. Now, any sensible person will run away from a wildfire as fast as possible (the same, I suspect, would be true of the Vitruvian hero as soon as the lava began to spit). But if someone comes back to the blackened forest a week later, when the ground has cooled off, it is not hard to find snags and stumps where coals are still glowing. What this means is that fire, in some form such as this, was ready for our ancestors' taking a great many million years before they were ready to take it.

At the same time, it now appears that our ancestors were indeed ready to take up the use of fire perhaps one million years earlier than had previously been supposed. Work at an archaeological site in East Africa, announced in November, 1981, has turned up solid evidence of fire in association with stone tools; it dates back to more than 1,420,000 years ago (give or take 70,000 years). The evidence, found by a quartet of scholars in Kenya (where else?) at a place called Chesowanja, consists of numerous bits of baked clay. Analysis indicates that the clay had been transformed from its natural state by exposure to a campfire rather than to wildfire. Whose fire? Probably *Homo erectus*'s.

It will be helpful if we stop from time to time to summarize man's progress in his quest for control over energy. For ease of comparison I will try to do this under uniform headings. Here is a good place to start:

TABLE ONE

PERIOD	NON-MUSCLE ENERGY AVAILABLE	MACHINES
Villefranchian Initial Paleolithic (Pleistocene up until 1.2 million years before the present, or B.P.)	FIRE	?
Donau/Günz through Riss I Classic Lower Paleolithic (Up until 200,000 B.P.)	FIRE	*Lever class:* digging stick *Wedge class:* hand ax

Two of these terms need explanation. "Before The Present" (abbreviated B.P.) is a phrase that came into use following World War II when sensitive methods of measuring long intervals of time were first developed. It soon became clear to workers in this field of "absolute" chronology that it was at least clumsy and silly to write about things happening 1,000,000 years or even 10,000 years before Christ as B.C. (the Western convention).

For example, the use of this pious benchmark forces everyone to mentally add some 2,000 years to a given date to determine its actual distance from today. Consider a rock sample dating back in traditional terms to 1,000,000 B.C. Such an object would actually be 1,002,000 years old in the year 2000 and would have been 1,001,981 years old in 1981. Thus a new benchmark, "the present," was introduced (defined as A.D. 1950). Those who still want to give dates in years before Christ are welcome to subtract 1,950 years (or, as here, a rounded 2,000 years) from the B.P. date.

As to the other term, "efficiency," I have defined this as the application of one tenth of one horsepower over a 24-hour

KNOWN MATERIAL CULTURE	NOT KNOWN BUT PROBABLE	EFFICIENCY 0.1 HP. x 12 hr. equals 50%
Simple pebble & flake stone tools, rarely made of flint.	Tools improvised from deadfall wood, animal bones, & horns. Carrying devices. Use of natural shelters.	minimum: average 50%
Acheulean core-tool tradition. First known wood artifact (ca. 250,000 B.P.). Use of natural shelters.	As above, plus: Fire-hardened sticks and spears. Carrying devices made of cordage and/or skins.	modest advance: average 65%

period. One horsepower is defined as the ability to raise 550 pounds to a height of one foot above the ground in one second, or to raise one pound 550 feet high in the same one-second interval (or any other foot-pound per second equivalent). Now, man's muscle power is just about equal to one tenth of a horsepower. If we take for granted that man does work at the rate of 0.1 horsepower during the annual average of 12 hours of daylight per day, then, before fire was available, his efficiency rating would be 50 percent (12 hours of work in every 24 hours). Because fire extended the illuminated period of the average day by 15 percent, we can raise man's efficiency rating to 65 percent as soon as we find evidence of fire-tending. The rate can be increased further when man begins to apply the muscle multipliers that the Greeks called the five "simple" machines, engages in cooperative food-collecting, and so forth. When man learns how to use animal muscle, in addition to his own, the efficiency rate passes the 100-percent level and becomes a meaningless measure. That point on our trail still lies far ahead.

CHAPTER
4

Applying Energy in the Old Stone Age

SILICON IS one of the most abundant elements; it makes up nearly 30 percent of the earth's crust. The form of silicon most familiar to us is glass: a mixture of sand (silicon dioxide) and other materials melted to the consistency of taffy and then, if it is to be window glass, poured out to settle and harden again on a flat surface. If you have ever (by accident, of course) shot a BB pellet at a windowpane, you will have noticed the peculiar shape of the hole made by the impact. On the side of the glass that was struck by the pellet the hole is no bigger around than the tiny pellet itself. On the other side, however, the impact has punched out a sizable cone of glass. If the glass is thick enough, the base of the missing cone can be as big around as a dime. The crater that the impact created is not perfectly smooth; it contains wavy hills and dales. Geologists have a name for this kind of wavy breakage. They call it conchoidal fracture, because the alternating ridges and depressions look something like the ridged surface of a conch shell.

Not only glass, but most of the rocks that are largely composed of silicon break in this conchoidal way when struck with skill. (Brute hammering merely shatters them.) The flake that is detached by a cunning blow has a razor-sharp edge. One such rock, obsidian, actually is a kind of glass — cooked in a volcano instead of in a glass foundry. Crystalline quartz and vein quartz are also siliceous rocks, and so is flint and its close cousin, chert. The latter two are often found in the form of irregular nodules embedded in the kind of soft sedimentary rock we call chalk.

We have seen that our forebears of two million or so years ago in East Africa produced the sharp flakes they used to cut up meat by hammering fine-grained volcanic rocks with water-worn pebbles, objects we call hammerstones. We have also seen that at Chou-kou-tien, the fire-tending toolmakers preferred vein quartz. In general, all around the world the makers of chipped stone tools have shown a similar preference for siliceous rocks. In Western Europe and the British Isles, where the record is best known, flint enjoyed overwhelming popularity. Bear in mind that the record is best known in Europe primarily because this is the part of the world where most of the studies of Old Stone Age archaeology have taken place. The greater part of the work, moreover, has been done by French archaeologists (perhaps as a kind of unconscious apology to old Boucher de Perthes).

Given a flint nodule and the desire to make a stone tool, two basic approaches are possible. You can remove what you don't want and keep the rest, or you can remove what you do want and forget the rest. If you do the first, you end up with what archaeologists call a core tool. If you do the second, you end up with one or more flake tools. You can also do both and end up with a tool kit that includes both flake and core tools. This is what the spearmakers at Clacton-on-Sea and the possible pyromaniacs at Hoxne did. A recent analysis of 246 flint artifacts — cores and flakes — from Clacton, showed that only 40 of the total, most of them flakes, had actually been used. Evidently the knappers of Clacton considered the other 200-odd flakes and cores somehow unsuitable. Of the 40 that were used, 38 were flake tools and two were core tools. This means that the utilization ratio for flakes was about 20 percent and for cores exactly 10 percent, which shows that the first students of the Clactonian flint industry were right when they said it was predominantly a flake industry. At Hoxne, a few score miles away in Suffolk, the contrary tradition was observed. The emphasis was on core

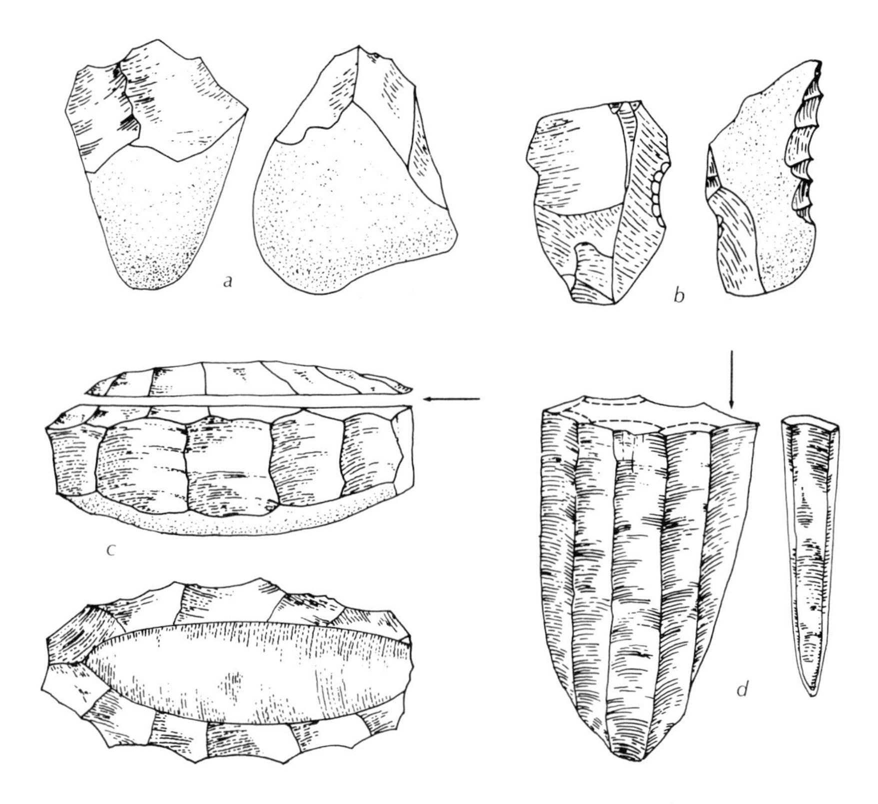

From crude to fine: simple pebble tools (*a*) were among man's earliest artifacts. Some from East Africa are probably more than a million years old. More extensively flaked tools made of flint or chert (*b*) were made later; these two could have been used for woodworking. By Middle Paleolithic times the flint knappers were thoroughly skilled craftsmen. Here (*c*) a core has been carefully trimmed in preparation for striking a fine flake from it, Levallois style (*arrow*). A top view of the core, below, shows the smooth conchoidal scar left behind by the detachment of the flake. By Upper Paleolithic times the knappers could detach fine razor-sharp blades all around the circumference of a prepared core (*d*).

tools, specifically the tidy bifacial *coupes-de-poing*, or handaxes, characteristic of the Acheulean industry. In making a biface, because you remove what you don't want, you end up with a lot of waste flakes. Indeed, a recent analysis of more than 400 flint artifacts from Hoxne showed that only 9 percent of the flakes in the collection had ever been used as tools.

What has all this to do with energy? The answer is simple. We have already seen how our ancestors improved their efficiency by taking up meat-eating, by extending the workday, and probably by hunting in groups. These actions were all, in effect, timesavers. Now we are about to see another kind of saving: conservation of materials.

The making of a hand ax wastes a lot of flint even when, as at Hoxne, some use is made of the leftover flakes. So? Isn't there flint to spare? Yes, there is, but someone still has to find a source of nodules and do his knapping there. Or he has to collect the nodules along a stream bank, where they have washed out of the chalk, and then carry them back to his home place. Or he has to dig them out of the chalk himself (some of Europe's earliest mines, by the way, were flint mines; they have been found all the way from Britain to Poland and Hungary). In any of these cases, a Clacton knapper has it all over a Hoxne knapper. One nodule in the Clacton man's hands will yield a lapful of flake tools whereas the Hoxne knapper ends up with only one hand ax. If you wish, you may note this down as the world's earliest economic statistic.

Volumes have been devoted to the artifacts of the Old Stone Age. Here we must go at a gallop. As we move along, however, note through the blur of passing time that man's flint tools are becoming better made, smaller, and much more uniform. The hand axes in particular look better and better. They also grow fewer in number and finally disappear. The flake tools, too, look less and less like kindergarten flower petals modeled in clay and more and more like assembly-line razor blades. Right now, in our gallop, we are passing through a suburb of Paris on the right bank of the Seine, Levallois-Perret. It was here that the use of prepared flint cores for the production of flakes was first recognized by archaeologists.

The Levalloisian knappers preshape a flint core until it is about as long and wide as they wish the flake to be; they also prepare what is called a striking platform, which is where the blow that will produce the desired flake is to fall. Levalloisian flakes struck from such prepared cores are as much as six inches long. The outer face shows the scars of preparation, but the inner face is the smooth product of a single conchoidal fracture.

The great contemporary French prehistorian, André Leroi-Gourhan, has calculated the length of cutting edge that can be manufactured from a two-pound flint nodule by various core-tool and flake-tool techniques. A nodule of that weight will yield only one Clactonian chopper, and that so rough in execution as to have a mere four inches of usable cutting edge. An Acheulean knapper might obtain two bifaces from the same two-pound nodule; the combined length of their more smoothly fashioned cutting edges would be 16 inches, a 300 percent increase. Leroi-Gourhan did not include Levalloisian flakes in his calculations so I have added the Seine-side knappers to his array on the basis of my own horseback estimate. If they could fashion three prepared

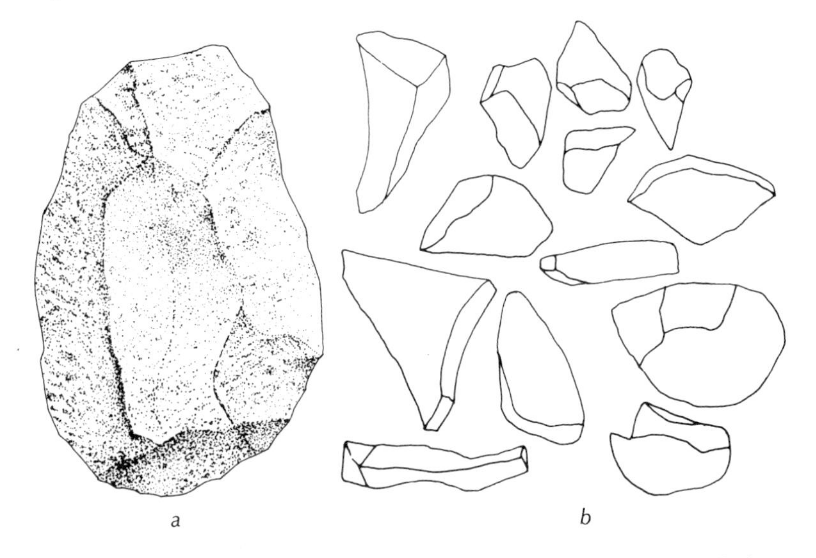

"Chopping tool" (*a*), made of basalt, is five inches long. All the flakes at the right (*b*) were struck off the basalt core in order to shape the final product, so that nearly as much of the rock was wasted as was eventually used. This is recent toolmaking, part of an experiment to see what the best uses were for tools such as this. The experimenters found that, far from being a good chopper, this tool was an excellent wedge for peeling off an antelope's hide.

cores from a two-pound nodule and strike a single flake from each, they would, by my reckoning, end up with 24 inches of cutting edge — 500 percent more than the Clactonian total.

If you will bear with a mangled metaphor, let us continue our gallop and still keep an ear cocked to Leroi-Gourhan's words of wisdom. We have now traveled the distance from Paris to Bordeaux and bridged a span of years stretching from the cold of

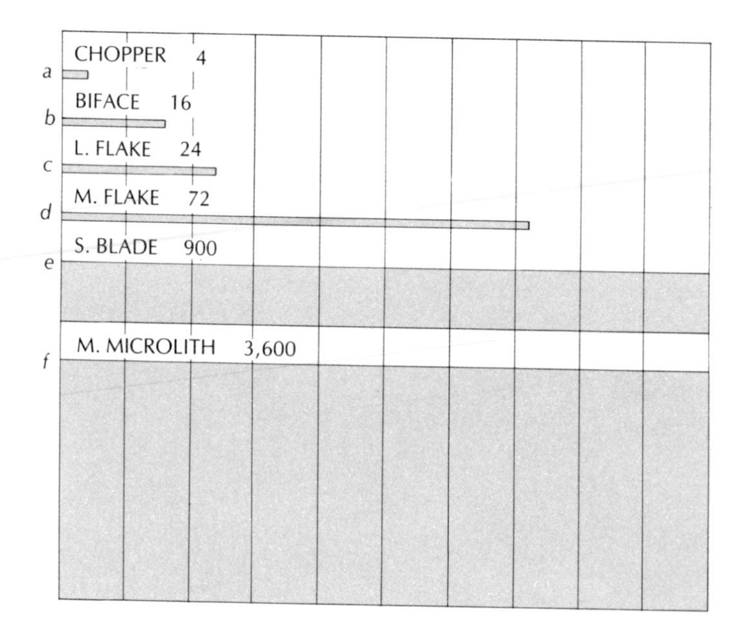

An early economic statistic: how much cutting edge per two-pound nodule of flint? For a single coarse chopper, say, maybe four usable inches (*a*). The knapper of Acheulean bifaces could make two of them from one nodule and end up with 16 inches of cutting edge (*b*). The knapper of Levalloisian flakes could prepare three cores, strike a flake from each and end up with 24 inches of cutting edge (*c*). The Mousterian knapper gets 72 inches from his two pounds (*d*) and the Solutrean maker of short blades gets 900 inches (*e*). Finally, the Magdalenian knapper of microliths can produce more than 3,600 inches of cutting edge (*f*) from his two-pound flint nodule.

the Riss glaciation to the temperate phase before the first onset of the Würm ice sheets. At Le Moustier, in the Dordogne region east of Bordeaux, the knappers are preparing cores that are smaller than the Parisian ones. They are also striking not just one, but a dozen or more flakes from each prepared core. The flakes are short but sturdy, and eminently suited for retouching into many different kinds of specialized tools. What has Leroi-Gourhan to say about this? The knappers at Le Moustier end up with a total of 72 inches of cutting edge per two-pound nodule of flint. From 4 inches to 72 inches is a great advance — 2,300 percent — but there is better still to come.

In terms of time, our gallop is now slowing down to a trot; in terms of space, we are backtracking from the Dordogne to a point some sixty miles northeast of Lyon. Here, near the village of Volgu, was unearthed a hoard of 17 flint blades shaped like laurel leaves, only very much larger. The largest of all was more than a foot long but only three-quarters of an inch wide at its widest point and scarcely more than a quarter of an inch thick! These masterpieces of Upper Paleolithic flint-knapping, produced some 20,000 years ago by artisans of the industry known as the Solutrean, will not again be equalled in artistry until Neolithic times in the Old World and well after the time of Christ in the New World (where the knappers worked principally in obsidian rather than flint).

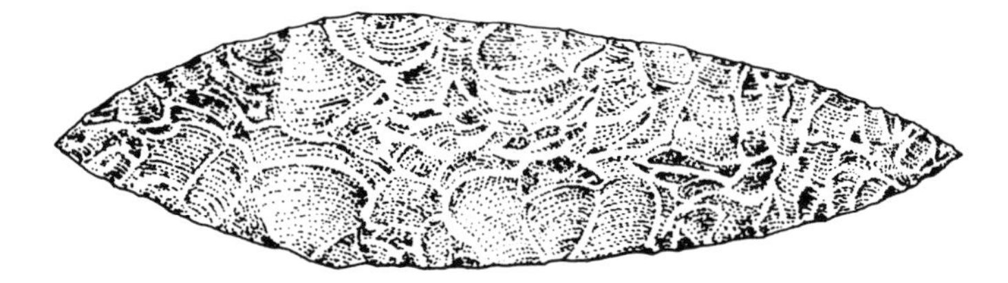

The ultimate in Paleolithic flint-working: a Solutrean "leaf" blade. The largest known is only about ¼ inch thick and ¾ inch wide but more than 12 inches long. The one seen here, 11 inches long, is less than a half inch thick. What they were used for no one knows; perhaps they were purely ceremonial knives.

The Solutrean laurel-leaf "blades," both those from Volgu and those discovered at other sites located from west of the Rhone to as far away as Spain, were surely not intended for day-to-day cutting or piercing. Indeed, they represent a complete reversal of the increasing economy in the use of raw materials that we are at such pains to observe here. The significance of that reversal and other similar "wasteful" investments of energy during Paleolithic times we will discuss later. Our main reason for pausing at Volgu is to point out the extreme sophistication of flint-working during the Upper Paleolithic. This sophistication, while evident in the elegance of the laurel-leaf points, was also applied practically in the production of standard flint blades from cores that were prepared in a new and extremely simple way. The Solutrean knapper, his Upper Paleolithic forebears and their descendants in the craft all over Europe simply broke their oval flint nodules in half, rested the rounded end on the ground, and struck off blades all around the circumference of the half-nodule, using the broken flat end as a striking platform.

In this way the knappers of the Upper Paleolithic almost literally unwound a continuous sheet of blades until the core had shrunk to so small a circumference as to be unworkable. The end product was, in effect, a series of uniform blanks, say two to three inches long and a half to three quarters of an inch wide, that the knapper could further work into any kind of cutting or piercing tool he wanted. Leroi-Gourhan estimates that two pounds of flint, handled in this fashion, would yield 900 inches (more than 22 yards) of cutting edge. The jump from 72 inches to 900 inches makes the earlier advances seem trivial.

At Volgu we are within 20,000 years of the present. If we now continue at a slow trot we end up, 5,000 years later, back in the Dordogne again, at a rock shelter known as La Madeleine. This is the type site for the final Upper Paleolithic industry in western Europe, the Magdalenian. Within 100 miles of this rock shelter are found most of the great examples of Paleolithic cave art. Our ancestors' investment of energy in these magnificent but hidden displays of artistic virtuosity is no easier to explain than are the actions of the Solutrean laurel-leaf knapper.

Here, however, our primary concern is with the utilitarian side of Magdalenian life and with Leroi-Gourhan's final statistic. The Magdalenian knappers followed the Upper Paleolithic blade-tool tradition but they prepared smaller cores and produced far smaller blades than those of the Solutrean industry. The yield from a two-pound nodule? Three thousand, six hundred-plus inches of cutting edge, or more than 100 yards! In this mangled metaphor of a journey we have spanned 235,000 years in time. If I may invent a further addled statistic to go with Leroi-Gourhan's scholarly estimates it is this: the approximate rate of human progress in the production of cutting edges is thus one added inch per 65-year lifetime.

Actually, this may not be quite as silly a figure as it first appears to be. Of course, no constant rate of progress was involved; the advances were in leaps and bounds instead. A probably more meaningful statistic comes from comparing the lengths of time between leaps. For example, only 5,000 years elapsed between the time when the knapping output was 900 inches of edge per two-pound nodule and the time when the output reached 3,600 inches. In contrast to this, the jump from 4 inches to 72 inches required 150,000 years. If we call this longer interval the equivalent of 100 percent, then the duration of the shorter interval is less than 4 percent of the longer. Perhaps we are seeing an example of what some people call the accelerating rate of history, already at work in this prehistoric setting.

The Holy Cross Mountains of Poland include among their rocks certain limestone beds that contain flint of a distinctive color: chocolate. This flint and the artifacts made from it are unmistakable. For this reason, Polish archaeologists have not only been able to locate the "mines" where hunters of the Upper Paleolithic dug out nodules of chocolate flint, but have also found the "factories" where the nodules were flaked into blades for use in cutting wood and meat. Moreover, studies of flints found at various ancient hunters' camps throughout Poland reveal how far artifacts made of chocolate flint traveled from the

Holy Cross mines. These findings, too, provide some of our earliest economic statistics.

The mines were simple pits; at the same time, the pits were big enough to represent a considerable labor investment in earth-moving. The pits at Oronsko, for example, were 4 to 6 feet in diameter and some reached a depth of 10 feet (to dig a pit 6 feet across and 10 feet deep requires the removal of 160 cubic feet of soil and rock — an estimated 28,000 foot-pound investment). The Polish archaeologist Romuald Schild has found that Upper Paleolithic campsites as far as 60 miles from the chocolate-flint mines contain artifacts made from chocolate flint in abundance; single pieces have been found as far away from the Holy Cross Mountains as 150 miles.

Are we seeing, in the digging and moving of chocolate flint, the first evidence of industry and trade? As regards the mining industry, the answer is no; there were earlier mines. What is probably the oldest mine in the world is in Africa; the product that was mined there 50,000 or so years ago was not flint but red ocher, an earthy form of iron ore much used as a pigment. As regards trade, the answer is also probably no, but with a difference. The hunters who inhabited the open forests of Poland in Upper Paleolithic times traveled considerable distances each year, following migrant game animals. Passing the Holy Cross Mountains they could have supplied themselves with a nodule or two of the prized chocolate flint; some of them could even have carried off a few spare nodules for barter. The pits could have been enlarged and deepened in successive seasons by successive generations of hunters. But this is a long way from a population of miners and formal networks of trade. These two developments we will encounter in Neolithic times, a little farther down the road.

In any event we are almost at the end of our concern with flint. Well before the last of the Upper Paleolithic period, some 15,000 years ago, compound tools — for example, artifacts that combined bone or antler with bits of flint — had been generally adopted. Flint would continue to be present in the toolkits of woodworkers and of workers in bone and antler (and in the artists' workboxes) for thousands of years more. It would be used by Neolithic artisans in Denmark to make imitation Bronze Age daggers and by Bronze Age artisans in Egypt to make exquisite ceremonial knives. Today the knappers of Brandon in England still produce fine flint blades that are sold to those who maintain flintlock firearms, while wheat farmers in Turkish villages still knap flints to set in the bottom of their threshing sleds as cutting edges. But all these are special uses. The age of flint was destined to end not long after the end of the Old Stone Age.

What did the hunter-artisans of the Ice Age use their flint tools for? This question has challenged the imaginations of prehistorians from de Perthes onward. The shapes were studied and a great naming began. "Hand ax" was one of the first names assigned to core tools presumably used for chopping. "Scraper" was one of the first names assigned to heavy flake tools. "Burin," the French name still given to a modern metal-engravers' tool, was assigned to flint flakes that had a sharp corner or two. So it went. Not until a few years ago, however, did anyone bother to make modern flint replicas of the Paleolithic forms, use the replicas for different kinds of work, and then compare the wear patterns of the ancient and modern artifacts.

One such experiment was conducted recently by a young prehistorian, Lawrence H. Keeley. He used modern flint replicas to cut wood, antler, and bone, to scrape fresh and dry hides, and to cut both meat and non-woody plant material. He found that each such usage produced a characteristic kind of wear on the edge and surface of his tools; the different "work polishes" were readily distinguishable under the microscope. Keeley then examined a number of implements from English Lower Paleolithic sites, also at high magnification. Among other things, he found that flake tools from the Essex site, Clacton-on-Sea, had been variously used for cutting meat, cutting wood, scraping hides, and, to a lesser extent, working on bone. One of the crude choppers from Clacton also showed recognizable wear; it had been used both to cut wood and to bore holes in wood.

Keeley went on to analyze flakes and hand axes from the Suffolk site of Hoxne. Two of these more refined core tools, he found, had been used to cut meat. The Hoxne flakes had served, like those from Clacton, to cut meat, wood, and bone, and to scrape hides. Indeed, some of the flakes known traditionally as side-scrapers and presumed to have been used in dressing skins actually showed the work polish that is produced by hide-scraping. Other Hoxne flake tools had been used to cut non-woody plant materials; perhaps the Hoxne hunters gathered ferns and reeds for bedding.

Using flint tools to butcher meat and to dress hides is direct use; so is collecting the materials for a soft bed by the fire. Cutting wood, antler, and bone, however, means using a basic tool to make a tool of another kind. We can forget about recovering the tools that were made of wood; finding of just one such at Clacton was greater good luck than anyone has a right to expect. As for the antler and bone artifacts, a good number of these have survived their millennia of burial. Those that are most abundant are from Upper Paleolithic times, and once again the French collections are the richest.

A common bone artifact is the gorge, or toggle: a short length of bone that is tapered to a point at each end. These could have been used as fishhooks (a kind of hook quite different from the curved and barbed hooks we know) or as clothing fasteners. Barbed lance heads were made of bone and of antler. Some are hollowed at the base and may have been hafted on a wooden shaft, as modern harpoons are. Much more ambitious are lengths of antler, carefully pierced with a sizable hole at the point where a tine branches off; the first French archaeologists to find these dubbed them *bâtons de commandement,* imagining them to be the Ice Age equivalents of the 18th-century French symbols of military authority. (Although it is impossible to be certain, today these odd objects are thought to have been used as a kind of lever for straightening crooked lances and projectile shafts.) Equally

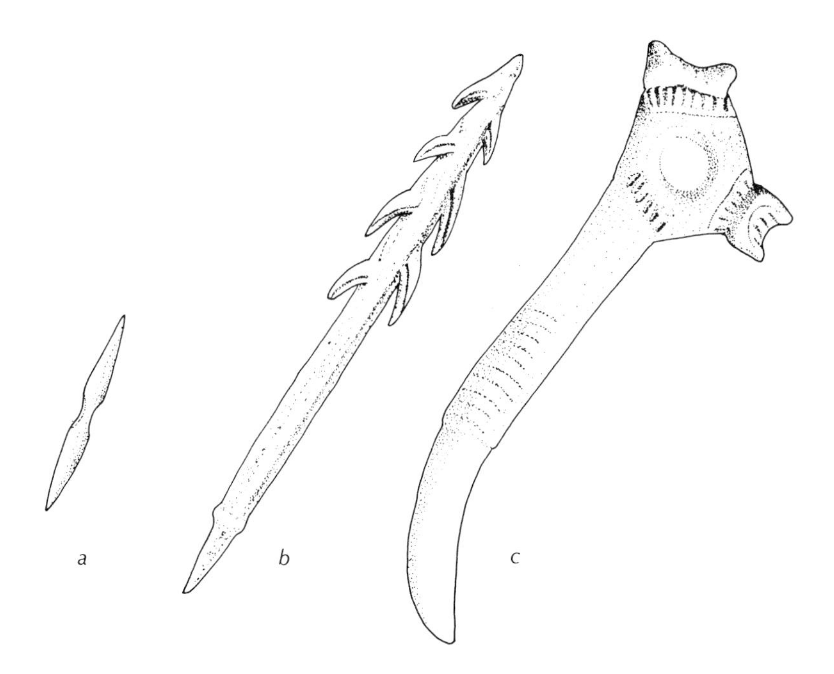

Some of the tools that men of the Upper Paleolithic made from bone and antler, using flakes of flint. The short piece with two points (*a*) is a toggle; it can be used as a kind of fishhook or as a kind of button. The longer piece with barbs at one end (*b*) could have been a lance head or a harpoon. The odd bit of antler with the hole bored through it (*c*) may have been used as a wrench for straightening crooked wood shafts. The romantic French call them command batons.

ambitious are a number of bone and antler hooks, many of them elaborately decorated, that must have formed the business ends of another kind of lever, the spear thrower. This is a device that allows the hunter to extend the length of his arm artificially and thereby apply much greater leverage when throwing a lance: greater leverage, greater velocity and greater range.

These are not our ancestors' first use of the lever. I have suggested that digging sticks were among mankind's earliest tools, and digging sticks, of course, are levers. The baton and the

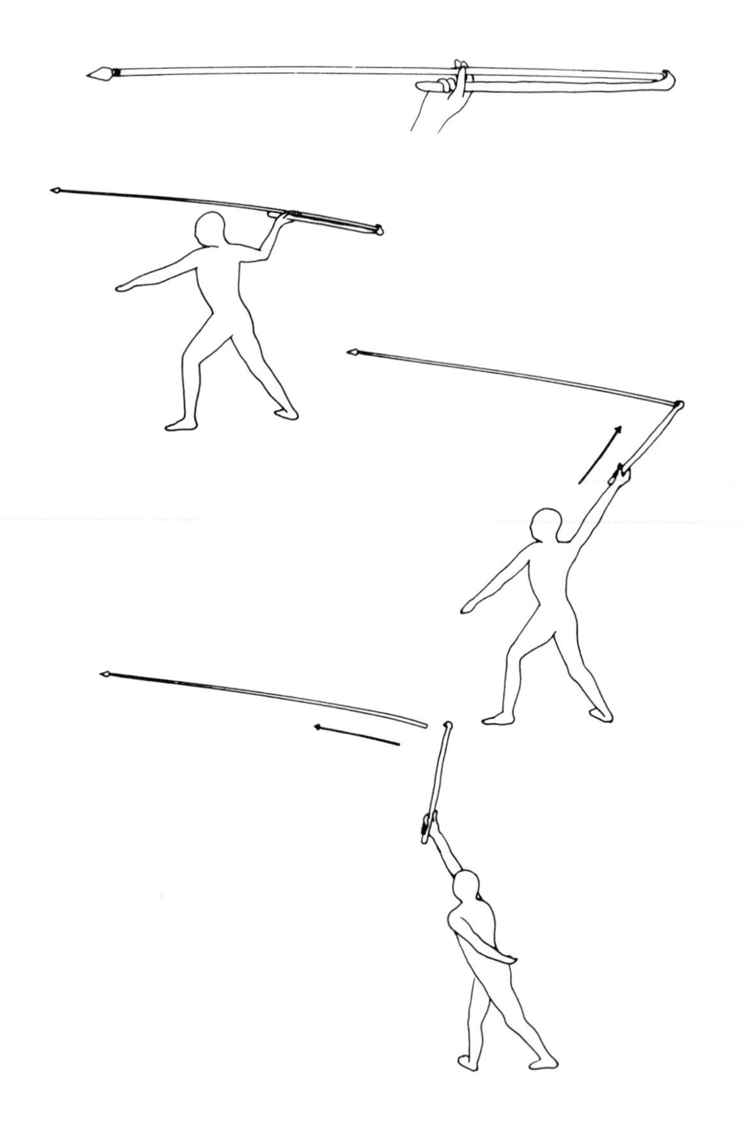

Add to the list of levers: the spear thrower serves as an arm extender and lets its user throw his missile faster and farther.

spear thrower, nonetheless, are new applications of the muscle-multiplying lever and, if the baton actually was used as a shaft-straightener, both implements served to improve hunting efficiency. Because we have been following the increased efficiency in flint technology so closely, we have gone ahead of ourselves regarding meat on the table. It is time to backtrack briefly and review the evidence of our ancestors' progress as hunters.

One way to assess this progress is to examine the animal bones found at various Ice Age campsites. It is the evidence of the animal bones, for example, that indicates a taste for venison among the *Homo erectus* families of Chou-kou-tien. Halfway round the world from China, hunters in Spain (whose stone tools represent the same Acheulean hand-ax industry that has been found at scores of sites elsewhere in Europe) were feeding on deer too. They also ate animals of more formidable size: wild horse, wild cattle, elephant, and rhinoceros.

Now, wild horse and wild cattle graze in bands, and bands of animals may be stampeded. This is a time-honored hunting technique. It was used by the early inhabitants of the New World to kill the buffalo, the wild cattle native to America, thousands of years before modern Europeans brought horses there. Stampeding was used so regularly in France during Upper Paleolithic times that the remains of wild horses at a single site, where the hunters' strategy was to stampede the herds over the edge of a cliff, have been counted in the thousands. Such herds may be startled and set on their way to destruction by loud noises, by the sight of approaching men, or by man-made smoke and flame (a method that it seems possible the Acheulean hunters of Hoxne may have used). All that the hunters require is organization.

One does not, however, stampede elephant or rhinoceros herds. In fact, I do not know of any primitive hunters in Africa or Asia who pursued the rhino as game before the invention of firearms let their pursuers slay these ill-tempered thunderers from afar. Daring men can hamstring an elephant, as the pigmies of the African forest still do on occasion, but this takes a certain knack and a sharp knife. I would not want to try it, even with a Solutrean laurel-leaf blade, let alone an Acheulean hand ax. Yet the fact remains that the bones of both mammoth and rhino are found at the Spanish Acheulean sites.

Does this mean that, like the hunters of East Africa more than a million years earlier, the Acheulean hunters simply scavenged the occasional elephant or rhino that they found dead of natural (or unnatural) causes? This is possible, but there is also another possibility. The hunters may have pit-trapped these very large animals. Once pinned in a pit, even an enraged elephant can be stunned with rocks and stabbed with lances until it bleeds to death. It is not so easy, thereafter, to butcher the pit-trapped prey, but it can be done. What I find harder to imagine is not leaving quite a lot behind, including a few accidentally dropped tools such as are found along with great quantities of bones at buffalo-kill sites in America. Nevertheless, pit-trapping is something Acheulean hunters might have done, although my confidence in saying so would be greatly increased by the discovery of even one such pit, complete with leftover bones and forgotten tools.

What makes Acheulean hunting different from the hunting and scavenging of East Africa is something quite aside from a greater variety of game animals and a more aggressive pursuit of prey. Even if no one dug pits, it seems impossible that these hunters could have stampeded game except by working together in more than family-sized groups. Let us speculate that a typical hunter's family consisted of an adult couple, one or two mature sub-adults, perhaps an infant, and possibly one or even two still vigorous grandparents. For driving game, such a family could form a beaters' line six strong at the most. But a drive requires not only beaters but also flank guards and a group of slaughterers at the end to dispatch the disabled prey. A band of five families made up of the numbers proposed here would provide just about the correct strength for a game-drive. A band of ten families would be better still, but a band any larger could result in overkill.

If hunting bands numbering five to ten families actually did exist among the Acheulean hunters, perhaps gathering together during the warm months and breaking up again come winter, then mankind had taken another major step forward. For groups can do things, besides driving game, that single families cannot. There is a place in a group for an older man or woman of declining strength who can still chip flint or make string out of bark just a little better than their fellows. These elders can be fed a share of what the hunters in the group bring back to camp and the cost is trivial; meanwhile such richly experienced persons can pass on to the youngest generation their accumulated skills and, just as important, their recollections of the past. The seeds that will in the future sprout into technician, teacher, and even priest are planted by the organization of group living.

You will note that I have used the term Acheulean hunter rather freely. Do not suppose I imply a population of hunters spread across Africa and Europe, all sharing a common culture and speaking some common language. All I mean is that, for reasons which may be explained in several different ways, the various human occupants of this part of the Old World and other parts far more distant all happened to make core tools that are remarkably similar in appearance. The human fossil remains are so scarce, however, that it is impossible to judge whether they were all men of the same evolutionary grade. For example, the *Homo erectus* population of Java, China, East Africa, and South Africa — and probably elsewhere in Eurasia — seems to have been such a single-grade group for a period of more than a million years. They did not however, achieve any notable cultural uniformity. In a word, there were no "Acheuleans" in the sense that we might say, today, there are Indo-Europeans (a group that shares a common ancestral language) or Buddhists (a group that shares a common ancestral religion). The fact that for 100,000 years or more our forebears were obsessive makers of hand axes by no means demands that they were also kissing cousins.

This picture of undemonstrable kinship changed when

Meat on the table. All eight of these mammals were depicted in the cave art of the Upper Paleolithic and all, possibly excepting the bear, were eaten. They are, from the top left, goat, pony, ox, and reindeer and, from the top right, horse, mammoth, bear, and rhinoceros.

the flint industry we call Mousterian appeared on the scene about 80,000 years ago. The stone tools of this industry — made, as you will remember, on sturdy flakes struck from prepared cores of flint — have been found in association with human fossils of the same physical type far too often for the association to be merely coincidental. The physical type is called Neanderthal, a name given to that originally maligned skullcap from the Feldhofen cave in the Neander Valley. With the arrival on the scene of a particular human subspecies — the formal name is *Homo sapiens neanderthalensis* — in association with a particular inventory of artifacts, we can, with the help of a little squinting sideways, peering now at the tools and now at the human bones, begin to sketch a somewhat more detailed image of the Paleolithic hunters who ranged the Old World from the Atlantic Ocean to the Caspian Sea and beyond toward the end of the Ice Age.

If the Neumanns of Düsseldorf and Bremen had not decided to classicize their family name (Neu = new, plus Mann = man) sometime in the 16th century, the cumbersome three-syllable word "Neander" (Greek *neos* = new, plus *anthropos* = man) might have remained blessedly rare. If, too, the former Neumann, Joachim Neander, the hymn writer, had been less prolific and admired during his short life, perhaps the fourth syllable, "thal," familiar to us all today, would never have been tacked on to the end of the classicized family name. Alas for history, we are stuck with this atrocious four-syllable marriage of Greek and German, for Joachim's ancestors did change their name and his admirers named a little valley near Düsseldorf in his honor. In that valley was Feldhofen cave where was found the human skullcap that received the name Neanderthal.

The Neanderthals were major actors in human prehistory for perhaps as long as 70,000 years, which is more than ten times the total span of recorded history. Their distinctive skulls and other bones have been unearthed not only in Germany but also in Belgium (the Spy pair are Neanderthals), in France (more than 30 of them), in Britain, Spain, and Italy, in Czechoslovakia, Hungary, and the Crimea (their houses have been found in the

Ukraine), in Palestine, in Iraq, and in Soviet Uzbekistan, less than 100 miles from Samarkand.

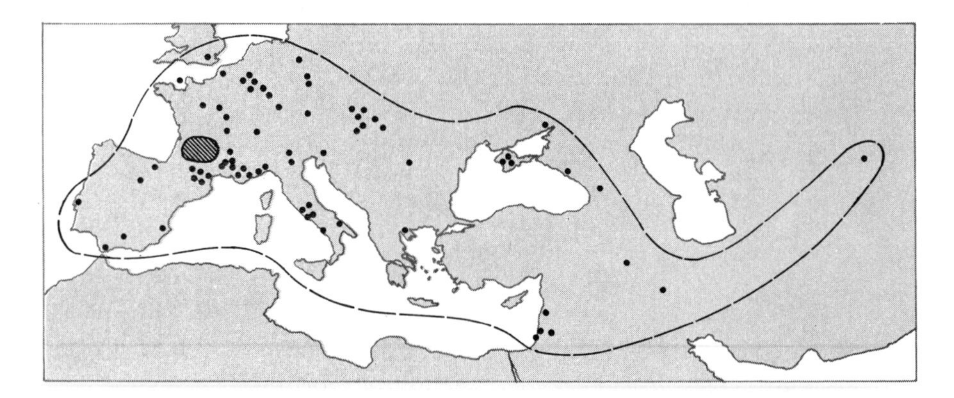

Homo sapiens neanderthalensis remains are most abundant in the western Massif Central of France (*hatched oval*) where at least 35 sites are known. Neanderthal fossils have also been found in many parts of Western and Eastern Europe, in the Levant, Iraq and Iran and as far east as Uzbekistan. If only Joachim Neander had been a less popular writer, this full-fledged member of the human race might instead have been named *Homo sapiens feldhofenensis*.

Physically, they have been slandered for decades as stooped, brutish lowbrows. This is the result of an accident not unlike the accident of their name. The first major study of Neanderthal physique was based on a relatively complete skeleton found in a French cave: La Chapelle-aux-Saints, near Corrèze in the Dordogne. The skeleton was that of a man, about 40 years old at his death, who suffered from extreme arthritis, senile degeneration of the neck vertebrae, and near-total loss of teeth. Chapelle, as we may call him, was indeed stooped, but it was his arthritis and not his genes that made him so.

Chapelle was also a lowbrow in the physical sense, but the Neanderthals as a group were not short on brains. The average human brain size today is 1,360 cubic centimeters; many Neanderthals had brains larger than 1,500 cc and at least one of them had a brain larger than 1,600 cc. Here, in fact, is the first evidence of a sizable population of big-brained humans in the entire fossil record. The only earlier representatives of man numerous enough to be considered a population, the *H. erectus* group that flourished as a genetically stable physical type for better than a million years, never managed to nudge their brain size significantly above 1,100 cc. With respect to the Neanderthal's alleged brutishness, a population that fed its old and infirm, buried its honored dead, and even once spread flowers on a grave needs no one to make apologies for it.

More to the point, in light of our specific interest in energetic efficiency, is the fact that two groups of Neanderthals, at least, may not have had to hoard the fire they captured in the wild. One evidence of this was found at the Cave of the Hyena, a Mousterian site in France. The excavators found a supply of a significant mineral, iron pyrites, among the Mousterian artifacts and animal remains buried there. The significance of the discovery takes a little explaining.

In our age of butane lighters and giveaway matches (matches were unknown before the 1820s), few give any thought to the problem of starting a fire. Indeed, this has been true ever since iron became a common metal, less than 4,000 years ago. Tinder, flint, and steel were the firemakers before matches came along. If we go farther back than that, we come to a time when essentially one had two choices when a fire was needed: spark or friction. For friction, you need a plank of soft wood and a rod of hard wood. You then rub the hard rod back and forth on the soft plank (or, more efficiently, spin the rod) until friction raises the temperature high enough to make the accumulation of fine duff produced by your rubbing (or spinning) begin to smolder. By blowing on the duff pile you can coax a hot glow that will light off whatever dry kindling you have ready. This is doing it the hard way, unless you have a bow-drill rig to make the rod spin.

Striking a spark is much easier. Two lumps of flint, hit together, will often produce a spark. By the 70th millennium B.C. our forebears had been banging at flint for perhaps one million years; they must certainly have been accustomed to seeing sparks fly. The only trouble is that the spark produced by two flints will not kindle a fire; it is only a kind of cold luminescence.

What our forebears needed was one flint and one rock of a different kind: a sulfurous stone. As it happens, the most common sulfurous stone is iron pyrites. Is it too much to imagine that, in two million-odd years of flint knapping, someone chanced to use a lump of iron pyrites as a hammerstone? If the

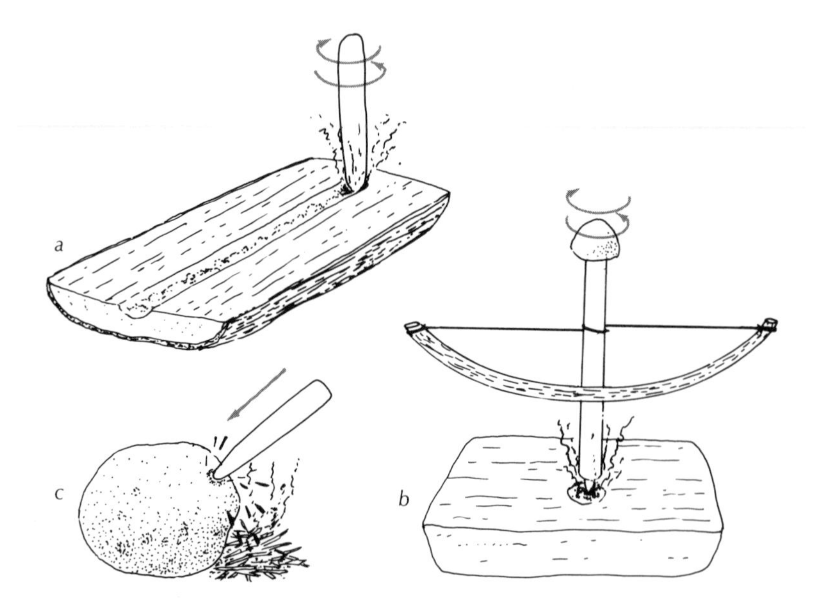

Starting a fire before the use of flint and steel. The hard way is with a hardwood rod and a softwood plank (*a*), rubbing the rod back and forth or twirling it between the palms of the hands. A better way (*b*) uses the hard-soft combination too, but a bow drill twirls the rod faster and your spare hand can put pressure on the rod, increasing the friction. A far better way (*c*) is to strike a special kind of stone, iron pyrites, with a piece of flint and catch the hot sparks on tinder.

resulting spark landed on the knapper's bare flesh, the difference between cold and hot sparks would have been learned in a twinkling. In any event, the finding of pyrites in the Cave of the Hyena strongly suggests that the Neanderthals there knew how to strike hot sparks. No earlier evidence of pyrites in man's possession is known.

What about friction? Evidence that such a fire-making method was known would, of course, consist only of a few bits of wood. As I have noted, this kind of evidence is not likely to have survived the passage of hundreds of thousands of years. One bit of wood, however, *has* survived the passage of some 100,000 years since Neanderthal man first appeared on the scene, equipped with his distinctive Mousterian kit of tools. The piece is a charred portion of a beechwood rod. It *may* be the remainder of a friction fire starter. The rod was unearthed at a rock shelter in Yugoslavia along with Mousterian tools. Is it the business end of a fire drill?

In these matters you have the choice of being conservative or radical. The conservative, thoughtfully fingering his beard, will say of the rod from Yugoslavia that it could just as well be the remnant of a fire-hardened stake or spear. He will also point out that the first-known nodule of pyrites that shows indisputable evidence of use as a striker stone is one unearthed from a Belgian site of Upper Paleolithic age. The Belgian nodule was used at least 40,000 years later than the pyrites bits from the Cave of the Hyena.

The radical, eyes rolling for effect, will cheerfully declare that evidence for man's ability to kindle fire at will so short a time ago as 100-odd millennia seems ridiculously late in the chronology of man's prehistoric progress. Both the friction principle and the spark principle could have been discovered many hundreds of millennia earlier. What is more, the radical will add, our pre-Mousterian forebears had every reason to seek an improvement in their fire-maintenance methods as soon as possible; their world was in the grip of a sequence of great cold periods. Either of these opposing views may be correct. For myself, neither conservative nor radical, I am content merely to hazard that some Neanderthal men knew how to *make* a fire when necessary.

Except for the bit of beechwood rod, not one other Neanderthal wood artifact has survived the 40-odd millennia of burial that separate the end of Mousterian times from the present. Some sharp-eyed workers in Palestine, however, have presented proof that the Neanderthals used a sharpened wood shaft offensively at least on one occasion. One of the Neanderthal skeletons excavated in the caves of Mount Carmel, near Acre in Palestine, showed a curious injury. A circular hole had been punched through the upper end of one thighbone and a pit appeared in the bone of the pelvis that lay behind it. It was clear that the individual had been hit hard in the hip with something sharp. The investigators greased the bones, lined up the hole with the pit, and filled the void with plaster. When the plaster hardened they found themselves with a cast of a smooth tapered point that could only have been made of wood. The poor devil had been speared!

Although more Neanderthal remains and Mousterian artifacts have been found in France than anywhere else, the most intriguing insights into the life of the Neanderthal come from Eastern Europe and Asia. In the Ukraine, the shelters they built are preserved as ground plans, along with surviving architectural elements. No Neanderthal bones have been found at the Ukrainian sites, but the stone artifacts are Mousterian, and Neanderthal remains are present not far away in the Crimea. Perhaps the most interesting of the Ukrainian sites are a cluster in the valley of the Dneister River: the Molodova sites, located southwest of Kiev. Excavators at the site known as Molodova I unearthed a scattering of mammoth skulls, jaws, tusks, and other bones that formed an oval enclosure some 30 feet wide in its narrowest dimension. Inside the enclosure were a total of 15 hearth areas. Reconstructing the kind of shelter this arrangement of bones and fires suggests, Soviet archaeologists envision a quite elaborate framework of upright and leaning wooden poles, roofed over with skins. The larger mammoth remains, in this reconstruction, were used to weigh down the edges of the skins where they reached the ground, while the smaller bones helped keep the skins pinned in place on top of the roof. Charcoal from the hearths, analyzed for its carbon isotope (C-14) contents, indicates that the shelter, probably a winter camp occupied for several successive years, was constructed about 45,000 years ago. The stone tools found on the living floor were Mousterian.

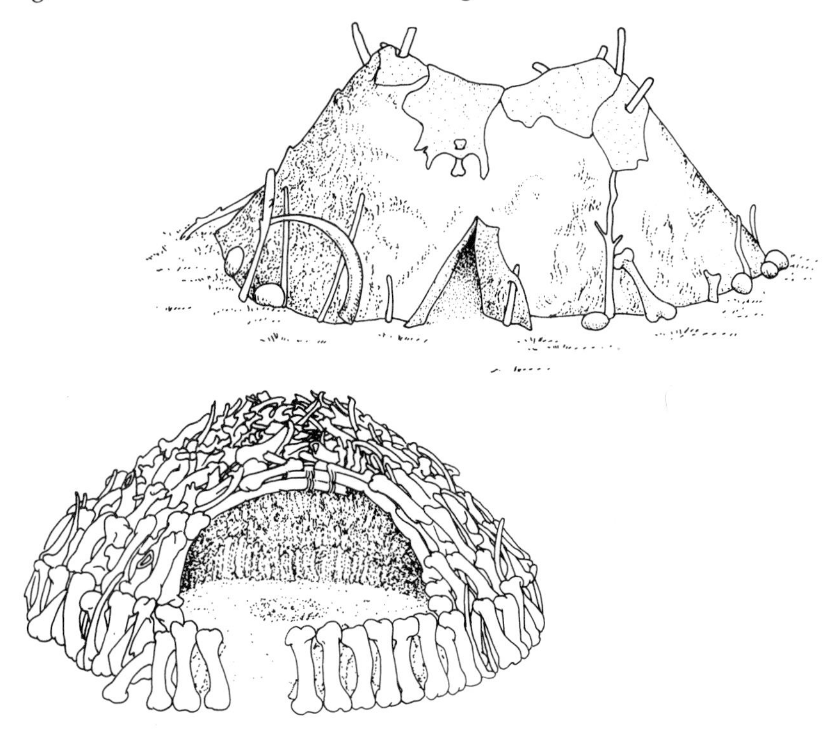

Winter shelters for the Ice Age hunters in the Ukraine (*top*) were lodge-like constructions of wood poles, presumably with skin coverings as shown here. Their Neanderthal inhabitants used bones and tusks of mammoths, deer antlers, and so forth as weights to keep the coverings in place. This has led to claims that the hunters built entire structures out of the remains of mammoths (*bottom*); this reconstruction was solemnly published in the U.S.S.R. Don't believe it.

Analysis of the animal bones found here and at other Molodova sites suggest that the principal game animal was the European bison. Deer and what are either wild goats or wild sheep were also eaten. Wolves, foxes, and hares were killed too, but evidently first of all for their fur rather than for their meat. Their skeletons were often found intact at one part of the site, but with the paws missing, whereas quantities of paw bones were found elsewhere. The fact that the small skeletons were not dismembered indicates that the Neanderthals did not eat the animals, or at least did not eat these particular ones. The fact that the paws were detached, the usual practice when skinning an animal for its pelt, suggests that the hunters, instead, were anxious to obtain the warm furs. If the Neanderthals did not sew the skins together to make winter clothing, they were not using their big brains. The key evidence of such activity, however, is the presence of bone needles and awls; none are found at Mousterian sites in the Ukraine or elsewhere. Indeed, almost no evidence exists for any Neanderthal use of bone or antler as raw material for tools.

The classic way to hunt bison, the most abundant of the Ukrainian prey, is to stampede them. Deer, sheep, and goats were more probably stalked and speared. The foxes and hares may have been clubbed and perhaps the wolves were too. What is certain, just to put your mind at ease, is that the enormous mammoths were not hunted at all. As evidence of this, chemical tests indicate that some mammoth bones at one winter camp were 1,000 years or more older than others. Further, a number of the mammoth bones show signs of having been gnawed and chewed on by carnivorous animals (in addition to foxes and wolves, the brown bear and the lion inhabited the Ukraine in those days). Finally, the mammoth bones at the winter camps are present in disproportionate numbers. There are more tusks and jaws than there are skulls to match them; vertebrae, ribs, and certain limb bones are absent altogether. The answer to this seeming enigma is obvious: the Mousterian builders could collect the picked-bare skeletons of long-deceased mammoths here and there on the upland steppe. The bones were as important a source of building materials as the ready-made lodge-poles available as windfalls in the pine forests of the river valleys.

The Neanderthal hunters of the Ukraine, then, fed mostly on bison but also ate deer and other hoofed game. If they tasted mammoth at all, it was probably by scavenging a few steaks from a freshly dead carcass. At another Neanderthal site in the U.S.S.R., Teshik-Tash, not far from Samarkand in Central Asia, the evidence indicates that the preferred game animal there was goat. Of the animal bones, 84 percent came from a single species of local wild goat. The Neanderthals of Teshik-Tash also ate horses when they could get them. Obviously these seeming dietary preferences have nothing to do with taste. It is simply that the mountains of Uzbekistan supported lots of wild goats and the Ukrainian steppe supported lots of bison. Let me note in passing that the only Neanderthal skeleton found at Teshik-Tash was that of a boy about nine years old. The skeleton was in a shallow grave, surrounded by five pairs of wild-goat horns. We will encounter other Neanderthal burial practices at the next site we examine.

In the Zagros Mountains, not far from Lake Urmia, the present-day borders of Turkey, Iraq, and Iran run together to form an inverted "Y." A little to the southwest of that intersection, on the flank of a local mountain, is a large cave that has sheltered man for some 100 millennia. This is the "big cave" of Shanidar; its mouth is 82 feet wide and 26 feet high at the center and the sheltered interior exceeds 10,000 square feet. Kurdish shepherds still corral their flocks in the big cave from November to April.

Between 1951 and 1960, Ralph Solecki, an American archaeologist, excavated this Iraq site to a depth of 45 feet. Beneath the layers of recent animal dung, Solecki and his colleagues successively encountered artifacts left behind by visitors from Neolithic and Upper Paleolithic times, including a Neolithic graveyard with 28 skeletons. Beginning about 16 feet below the surface, they encountered stone tools of the Mousterian kind and the bones of at least seven Neanderthal adults and two infants.

Analysis of the soil surrounding one of the seven adults, identified by the excavators as Shanidar IV, indicates that the body had received a kind of formal burial. If the seasons some 50,000 years ago were the same as they are today, Shanidar IV died sometime between late May and early July. A bed of boughs was prepared for the corpse and various flowering plants that were then in bloom — grape hyacinth and hollyhock among them — were strewn about the bed. What may have taken place thereafter the archaeological record does not reveal. To judge from the evidences of fire and the bones of animals found in association with some of the other Neanderthal cadavers at Shanidar, it is possible that the interring of Shanidar IV may also have provided the occasion for a ritual meal, but there is nothing to support this speculation.

So here we have a funeral service somewhat more to the modern taste than the one that saw the lad at Teshik-Tash buried with an array of goat horns. Nor is this the only evidence of, shall we say, sociality, found among the Neanderthal visitors to the Shanidar shelter. The adult identified as Shanidar I, who was evidently killed by an accidental rockfall, was a cripple from birth. His right scapula, or shoulder blade, and humerus, or upper armbone, were both underdeveloped. Both his radius and ulna, the lower armbones, were missing; this suggests the possibility that the useless lower arm may actually have been amputated earlier in life. Yet Shanidar I had lived to reach adult estate and his teeth show signs of heavy wear, which suggests that food was not stinted (and perhaps also, by analogy with Eskimo practices, that he may have been an active leatherworker). At least at Shanidar, evidently, to be a cripple was not a fatal liability.

Now, burying the dead is by no means a universal practice, even today. The Plains Indians of North America, until they lost their freedom, used to expose their dead on head-high raised platforms. Should you think this a barbarous custom, consider the Parsi of the Middle East, a highly cultured group that still practices Zoroastrianism, one of the world's more advanced religions. They honor their dead by carrying them to the top of special tall structures — Towers of Silence — where carrion-feeding birds dispose of mere mortal flesh.

Others prefer cremation, notably the Hindus and many Westerners who are tired of ritual that chiefly enriches the professional disposers of the dead. So did many inhabitants of Bronze Age Europe, to the intense irritation of the anthropologists, who are now forced to reconstruct the cultures without any idea of what the culture-carriers looked like. As we have seen, the Neanderthal practiced burial, at least on occasion.

In addition to the lad of Teshik-Tash and No. IV (and perhaps three or four others) at Shanidar, excavators have come upon some 24 examples of Neanderthal "manipulation" of the dead. Of these, 20-odd were all found at a single site in Yugoslavia; one was in Palestine, one in Italy, and two in France. At Mount Carmel the jawbone of a boar was placed in the shallow grave. At Monte Circeo in Italy, the burial consisted of a skull alone, but it lay within a stone circle and the base of the skull had been broken open after death. At La Chapelle-aux-Saints and Le Moustier in France the graves contained animal bones and stone tools.

The broken base of the Monte Circeo skull presents a puzzle that is multiplied in spectacular fashion by the findings at the Yugoslavian site — the same locale, by the way, where the charred beechwood rod was unearthed. The skulls of the 20-odd individuals of both sexes and various ages that were discovered there seem to have undergone brain extraction of a particularly complex kind. In each instance the fragile nasal portion of the face had been broken open, thus affording a narrow passage into the brain case (the brains of Egyptian mummies, many thousands of years later, were removed through the same aperture). Presumably the conductors of this grisly ritual picked out the brains through the nasal opening.

One might dismiss this reading of the evidence as the

excavator's own fantasy except that the larger limb bones of the dead — humerus and femur — had been split lengthwise, just as one would do to extract the marrow. The image this conjures up is that of a cannibal feast of brains and marrow and who-knows-what flesh. What, exactly, was going on?

When someone dies, the key problem is hygiene. The corpse will soon rot. If there is any universal practice concerning the dead it is that disposal is mandatory. At one extreme, the rot is stopped; ancient Egyptian artistocrats were embalmed to assure this, and so were Lenin and Mao Tse-tung. At the other extreme, the rotting are abandoned; primitive hunters, for example, will leave a corpse where it lies and flee any campsite where death takes place. If one looks at the Neanderthals' manipulation of their dead in this light, it might be conjectured that where no evidence of burial is found the corpses were simply abandoned, whereas those corpses that show evidence of manipulation were put underground for reasons of hygiene.

This is all very well, but it suggests no answer to the question of why certain burials include evidence of ritual that is simi-

TABLE TWO

PERIOD	NON-MUSCLE ENERGY AVAILABLE	MACHINES
Donau/Günz through Riss I Classic Lower Paleolithic (Up until 200,000 B.P.)	FIRE	*Lever class:* digging stick *Wedge class:* hand ax
Riss II through Würm II Middle Paleolithic (Up until 40,000 B.P.)	FIRE	*Lever class:* digging stick *Wedge class:* hand ax

lar in kind, if not in degree, to the funeral pomp of ancient Egypt. Nor does it approach being an explanation for nose-broken skulls and split limb bones. When we reach Egyptian times it is easy enough to explain the rich grave furnishings as prudent preparation for a future life of the spirit. Further, tales are told both in myth and in history of those who ate the flesh of heroes in order to become brave and others who drank out of cups fashioned from the skullcaps of their enemies. The truth of the matter is, however, that we do not know what the Neanderthals of Yugoslavia were up to. Nor are we likely to have much of an idea until many more sites of this period are discovered — and perhaps not even then.

One conclusion is nevertheless clear. Whatever belief and ritual underlay the Neanderthal manipulations of their dead, the fact that any manipulation took place at all means that this first population of large-brained humans had managed to deal with their environment efficiently enough to have won considerable spare time. Perhaps this is an appropriate place to summarize their progress. (For the sake of continuity an abbreviated version of the earlier summary is listed first.)

KNOWN MATERIAL CULTURE	NOT KNOWN BUT PROBABLE	EFFICIENCY 0.1 HP. x 12 hr. equals 50%
Acheulean core-tool tradition. First known wood artifact (ca. 250,000 B.P.). Use of natural shelters.	Fire-hardened sticks and spears. Carrying devices made of cordage and/or skins.	modest advance: average 65%
Mousterian flake-tool tradition (after ca. 100,000 B.P.). Fire-making kits. Use of man-made shelters (after ca. 60,000 B.P.). Wood lances. Ceremonial burials.	As above, plus: Dressed skins for clothing. Compound tools. Bone needles.	on evidence of shelter construction and time devoted to ritual, increase to 70%

Unlike the previous summary, this assessment of energetic progress during the interval from 200,000 to 40,000 years ago requires little by way of explanation. Perhaps one point is worth making. I put down as "probable" the existence of skin garments on the basis of the fox-paw and rabbit-paw finds in the Ukraine. What other use could be made of the pelts? This conclusion forces me to add bone needles to the "probable" list because the only way one can produce a garment from a number of small pelts is by sewing. Concrete evidence of skin garments and bone needles, however, does not appear until well after 40,000 B.P.

The second advance of the final (Würm) glaciation in Europe began to withdraw some time before 40,000 B.P. and a 10,000-year-long interval of mild weather followed. By the time the ice sheets began their final advance in about 30,000 B.P., the Mousterian technique of making tools out of flint flakes had gone out of style, to be replaced by the more refined kinds of pressure flaking and blade production already described as characteristic of Upper Paleolithic stone industries. Because we have equated the Mousterian stone industry with Neanderthal man, does the disappearance of these tools indicate the extinction of their makers? Such was the popular belief in the days, not so long ago, when not very many Neanderthal remains were known and the poor arthritic elder of La Chapelle-aux-Saints was being described as a typical example of a downright subhuman lineage. The facts that are available, however, can certainly be interpreted in another way. Men who were anatomically less robust than the Neanderthals evidently entered western Asia and Europe during the 10,000-year-long warm-weather prelude to the final Würm advance. (They, like ourselves, also had smaller brains but higher brows than the Neanderthals.) The archaeological record cannot yet say where these "modern" men came from or what they had been doing before their appearance on the scene in such areas as Iraq and the Ukraine. The same sites in Palestine that yielded the Neanderthal boar-jaw grave offering, however, contain human remains that suggest what the newcomers did upon arriving on the scene. Evidently they interbred; many of the physical characteristics of the Palestine hunters indicate a fusion of Neanderthal with "modern" genes. It is possibly this cheery evidence of romance at the end of the Middle Paleolithic and not the specter of extinction that accounts for the disappearance of one stone-tool industry and the appearance of a better one. In any event, the pace of prehistory quickened thereafter.

CHAPTER
5

Old Stone Age Energy Storage

I HAVE already mentioned that the Greeks, the earliest scholars whose understanding and analysis of technology have come down to us in written form, classified five specific devices as "simple machines." Hero of Alexandria, a man we will come to know better by and by, characterized the five as those "by the use of which a given weight is moved by a given force. Although they differ considerably in external form," he wrote, "they are all reducible to a single principle."

The five are, in order of complexity, the lever, the wedge, the endless screw, the wheel and axle, and the pulley. It is now time for us to consider the first of the simple machines, the lever, in some detail. For without benefit of learned analysis man had probably, by the beginning of Upper Paleolithic times, been using one kind of lever — the digging stick — for well over a million years. He was now to begin using a second kind of lever: the spear thrower. Once invented, the spear thrower would enjoy instant popularity. It came to be used worldwide (probably by reinvention) and was not abandoned until recently. For example, spear throwers were standard equipment in Mexico (the Aztec word for the device is *atlatl*) until well after the Spanish Conquest and were common in the New World Arctic until the Eskimo acquired firearms. Today spear throwers may even still be used by the Australian aborigines living in remote parts of the outback when they run low on cartridges.

Just how fast a spear can be thrown (and consequently

how hard it will hit and how far it will travel) is a simple mechanical formula: the available musclepower times the leverage applied. The standard leverage is one arm's length. If you have a stick equipped with a handgrip at one end and a hook at the other, you can fit the hook into a dimple at the butt of the spear shaft and grasp the handgrip in your clenched hand. The stick and the spear shaft both rest on the palm of your hand (you will have to spare a finger or two of the clenched hand to hold the spear shaft flat and firmly against the stick). You now raise your arm as if to throw a ball hard overhand. The throwing motion is actually the same as with a ball, but while you are reaching the top of the throw the hooked stick will rise up as if hinged to your clenched hand. Now the combined length of your arm and the stick together makes up the lever propelling the spear. The muscle power of your arm is no less and no more than usual, but it has been multiplied by added length, so that the leverage is far greater. How far greater depends on the length of the stick, your spear thrower.

Today, about the only contact we have with spears (and their shorter variants, lances and darts) is the javelin throw in track and field events. Purely a distance competition, the javelin throw combines unamplified musclepower with a high-altitude trajectory to obtain maximum range. The professional spear-users of the past, however, took pride in combining distance with accuracy. The spear in its simplest form is nothing more than a straight stick pointed at one end. With equipment no more complex than this, the aborigines of Tasmania (until we exterminated them) were able to throw a spear for a distance of 40 yards with such accuracy that the shaft usually passed through the bull's-eye of the target (a knothole in a plank). Forty yards is about the maximum effective range of a modern shotgun. A skilled hunter can come that close to a grazing animal by careful stalking; hunters during the Upper Paleolithic, armed with spearthrowers, had the option of trading off greater range against lesser shock at impact. It is not difficult to imagine that the strategy toward some kinds of game favored maximum shock (and penetration) and toward other kinds maximum range (and personal security).

Archaeologists at work in Europe (in France in particular) have unearthed a total of 66 Upper Paleolithic spear throwers. Most of them were found in southwestern France in strata identified with the final period of the Upper Paleolithic: the Magda-

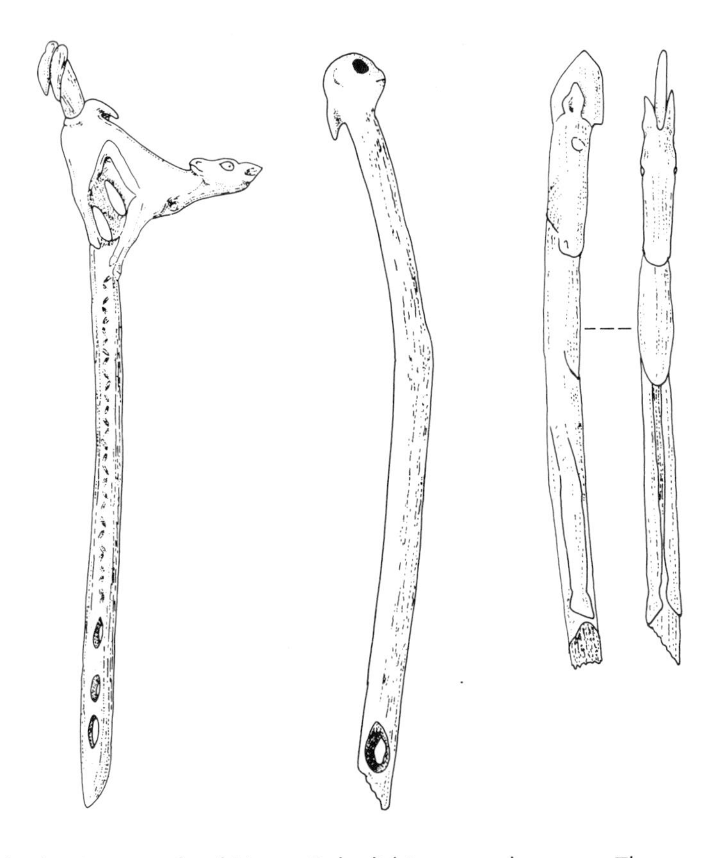

The business ends of Upper Paleolithic spear throwers. They may have been the work of specialists in bone and antler carving, for several hook designs, like that of the frightened ibex and the bird (*left*), were repeated. Indeed, 21 hooks that feature horses' heads (*right*) look enough alike to have been the work of a single craftsman.

lenian (named after a site we have already visited, La Madeleine, in the Dordogne). Most of the spear throwers are broken.

Of the 66, all but two were carved from reindeer antler; the exceptions, both broken, were carved from mammoth ivory. It is evident that many of the carvings were simply the business ends of spear throwers: short lengths of antler designed to be hafted to a wooden shaft. The business end was the hook end; no handgrips have come to light. The concept of a compound spear thrower — a wooden shaft with a hafted antler or ivory hook — suggests the possibility that most Upper Paleolithic spear throwers were made entirely of wood and have not survived their long burial. This possibility is strengthened by the fact that many of the hooks that did survive burial are decoratively sculptured. They may not have been intended for day-to-day use at all.

An inventory of the decorations — some of them incised outline drawings and others animal figures sculptured in the round — offers a potential insight into Upper Paleolithic hunting traditions. The horse is the favorite subject by a very wide margin: the total is 29. The next most numerous subject is the ibex: a total of seven. Next is the bison: five. Following this are three deer, two reindeer, one mammoth, one musk ox, an unidentifiable member of the cat family (lion or leopard?), and miscellaneous birds and fishes. Now, no one is likely to try to down birds with a spear thrower, nor is there any imaginable advantage in throwing a fish spear farther than one can see the fish. The spearing of big cats, too, is an unlikely pastime, although for different reasons — the foremost being personal safety. If this line of reasoning is valid, we can conclude that the spear-thrower decorations are not necessarily indications of their maker's preferences in game.

Game preference is a concept very much favored by the romantic interpreters of "Stone Age Art." They believe the painters and sculptors of Upper Paleolithic times were making magic with their paintings and sculptures in order to assure an abundance of the desired kinds of game when they hunted and even to promote hunting success. Why the artists chose various subjects remains, in my opinion, very much an open question.

The English archaeologist Dorothy Garrod, a shining light among students of the Paleolithic, has analyzed the total inventory of spear throwers. The distribution of similarly decorated examples she found suggestive of a certain widespread uniformity of what one could call Magdalenian "culture." For example, of the 29 spear throwers decorated with representations of horses, 21 look much the same: the end of the shaft has been carved to resemble the muzzle and mane of a horse's head, a forelock of the mane projects to form the hook that engages the butt of the spear. Sixteen of the horse-head carvings were found in the Dordogne, seven of them in a single hoard. The other five come from Kesslerloch, near Lake Constance in the Alps, some 450 miles from the Dordogne.

Garrod interprets this finding as evidence that a group from the Dordogne carried Magdalenian culture to the Alps. Considering how such prized materials as the chocolate flint of Poland came to be widely distributed, and giving added weight for what appears to have been a kind of Dordogne factory for the manufacture of horse-head carvings, I prefer to think that trade, rather than emigration, is a more likely explanation of the Kesslerloch find. In any event, the 21 horse-head carvings can scarcely be attributed to a similar inspiration suddenly striking 21 individual carvers.

The figures decorating three other Dordogne spear throwers present a nearly identical and complex scene, sculpted in the round. An ibex is precariously balanced, all four feet drawn close together. Its head looks backward; from beneath its tail an exaggeratedly large dropping is being expelled. A pair of narrow, pointed forms (birds?) are perched on the dropping; the pointed end of one of these is the hook that engages the spear butt.

Looking back today across a gap of more than 10,000 years, what are we to make of the sculptors' intent? The observation that animals often empty their bowels when startled must be as old as the art of hunting. This much of the composition, then, could be naturalistic: see what happens when you ambush an ibex. Some birds feed on dung too, but rarely while the dung is still in motion. The placement of the birds, then, if indeed these shapes are meant for birds, is at the least bizarre. Is this some kind of earthy Ice Age joke? Here again, as with the manipulation of the Neanderthal dead, the opportunity for insight is tantalizing but the evidence, alas, is mute.

If the spear thrower can be classified as a kind of lever, how shall we classify the bolas? You may find the word unfamiliar, but think for a minute. What does a Gaucho look like? Funny pants, with a sash at the waist; an odd hat. What's that thing wound around the sash? Some ropes with little leather bags at the end: a bola! Wrong — not one bola, two bolas, but one *bolas*. Right — usually three cords, knotted together, with a weight at the free end of each cord. The Indians of South America used them before the Gauchos came; the Gauchos quickly adopted the weapon. Some Eskimos also used the bolas before they discovered the .22 rifle.

You use a bolas by grasping the knotted end in your hand and swinging the weights round and round over your head. With a fairly small investment of musclepower you will soon have the weights traveling fast enough to make the cords hum in the air. When you are close enough to the animal you want to capture (the Indians hunted rheas, ostrich-like flightless birds, this way), you let go of the bolas so that it pinwheels through the air toward your target. If you've done it right, one or more of the cords will strike a leg or legs of the animal and the momentum of the weights will cause the cords to wrap themselves around the obstacle. At the least this will trip the animal up; at the most it will entangle its legs so it cannot move at all.

As you will have recognized, the momentum that causes the cords to trip the animal is nothing but our old friend, kinetic energy, in a rather unusual guise. By whirling the bolas around your head you have transformed potential energy into kinetic energy just as surely as if you had set a pendulum in motion or pushed a rock off the edge of a cliff. Moreover, the transformation is continuous; in a sense it is being stored up in the spinning weights. When you let go of the bolas you launch an independent energetic system. Those stones are loaded. Their kinetic energy will not be converted back into potential energy until they have stopped moving.

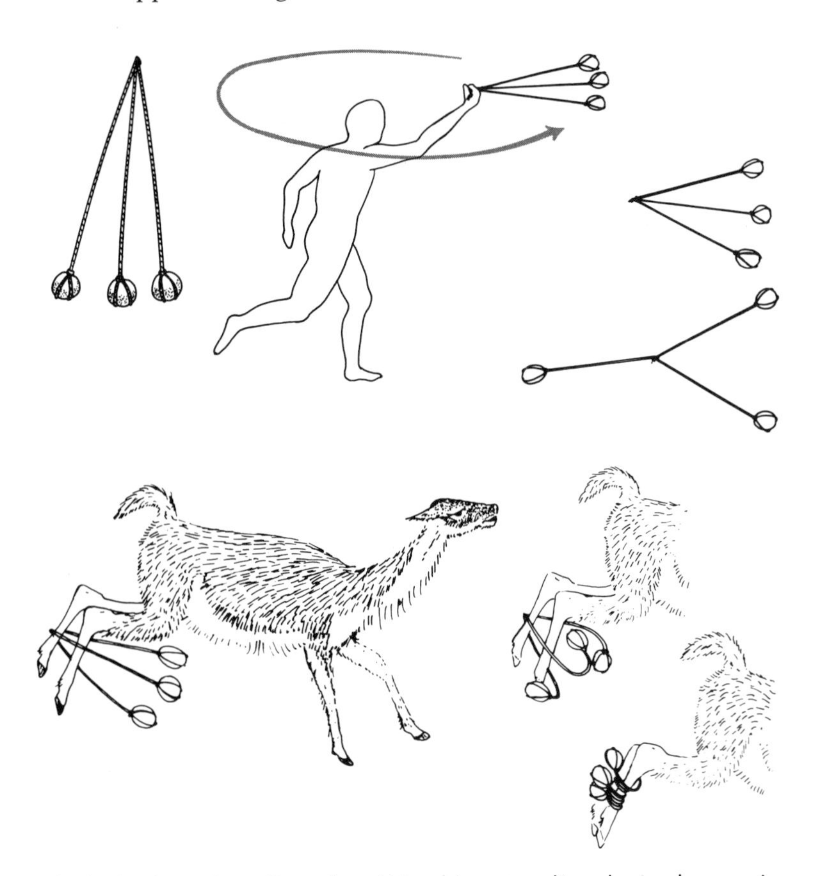

The bolas in action. Question: Was this entangling device known in the Old Stone Age? Stones that might have been bolas weights are known from that period, but no one can say whether they were used then as they were later in the New World.

So what does this weapon of the pampas and the Arctic have to do with the Stone Age? Both in East Africa and in Algeria stone balls have been found that had been laboriously chipped into spherical form. Sometimes they turned up in groups of three, sometimes in larger numbers, and sometimes in shapes other than perfectly round. Are these ancient bolas-weights? There is no way of telling. No such round stones have been found in Upper Paleolithic sites, but this bit of negative evidence is not necessarily damning. Even in Mousterian times, if a hunter wanted to make a bolas, a leather pouch filled with sand would make a perfectly serviceable substitute for a round stone. For that matter, so would a stone of irregular shape. The pouch would not survive burial except under most unusual circumstances, however, and irregular stones would attract no attention even if found in groups of three.

So we may admit the possibility that the bolas was a hunting weapon long before Upper Paleolithic times and even suggest the less probable possibility that its use in Middle and Upper Paleolithic times has somehow escaped documentation. Why bother? Because, if this is true, man had added to his energetic repertory an important item in addition to and very different from the Greeks' simple machines: a kind of storage battery. By whirling the bolas you are storing up kinetic energy at a modest cost in musclepower. When you let fly, the energy stored in the system begins to do work. So perhaps the bolas was man's earliest storage battery. We will not meet the second — the bow — until the Upper Paleolithic has drawn almost to its end.

Our Upper Paleolithic hunters made many things besides decorated spearthrowers. They pecked away on opposite sides of a small cobble until the two vee-shaped depressions met to form a hole. This finished product was probably mounted on the wooden shaft of a digging stick in order to increase the efficiency of this second kind of lever in their possession. They used another of the five simple machines, the wedge, to chisel bone into scraps suitable for needles. They made special projectile points, armed with barbs, from bone and antler. They made bone toggles — antique precursors of the button — to keep their garments closed. They pecked shallow hollows in stones (and collected stones with suitable hollows) and used them as lamps, with animal fat for fuel and a wick made of moss. They carved female figurines by the dozen, usually nude and far from naturalistic. Most of them seem grossly fat to our eyes but a few are slim enough to be cover girls, and one, from Ma'alta, a Siberian site, is clad in a close-fitting fur garment. A more representational sculpture, the head of a youth carved from an inch-high bit of ivory and unearthed in France, would be admired in any modern gallery.

They also made the first pottery. No pots, but pottery all the same. This will take a minute to explain. As you can guess, the vital prerequisite to the invention of pottery is a knowledge of the special properties of clay: first, that it can be molded and

Old Stone Age clay-play. Where caves contained natural clay deposits the artists of the Upper Paleolithic often amused themselves with the plastic material. One of the more ambitious clay sculptures consists of this pair of bison at Le Tuc d'Audobert in France.

modeled when damp; second, that it will harden permanently if kept hot enough for long enough. Upper Paleolithic man had learned both facts. As to modeling, where clay was present in the deep caves of the Dordogne, men spread their fingers wide on the cave wall to make long, wiggly parallel tracks in the clay. They even modeled low-relief figures of bison and a full-scale figure of a bear.

As to firing clay, we have to move from France to a mammoth-hunters' site in Moravia. There, one of the typical female nudes was sculpted, using a mixture of clay, fat, and bone ash. The figure was allowed to dry, some decorative details were incised on it, and then it was fired until permantly hard. Now, one such object could be dismissed as being the result of an accident. It might have fallen into a campfire, unnoticed. But the mammoth hunters did not stop after sculpting the Venus of Vestoniĉe. They produced miniature heads and full figures of bears, the figure of a mammoth, heads of reindeer, the head of a rhinoceros, and a number of lesser animal figures. Every one was fired to hardness. It is difficult to imagine selective accidental firings on a scale as large as this. Here, in what is modern Czechoslovakia, perhaps 10,000 years before the first Neolithic clay pots were modeled, are man's earliest pottery objects.

Thus far I have said a lot about saving energy but almost nothing about storing it. Nearly the only long-term means of energy storage open to man in Upper Paleolithic times was food storage. Tangible evidence of such practices is limited and indirect. The abundance and variety of the Upper Paleolithic diet, however, is evidenced by the animal bones found at the shelters. It would be surprising indeed if a people faced with such rich resources did not exploit them to the greatest possible extent. The saving ways that gave the men of the Upper Paleolithic the leisure to expand their artistic endeavors must surely have included food preservation.

What did the people find to eat during the mild interstadial and the last ice advance that followed? For one, fresh-killed

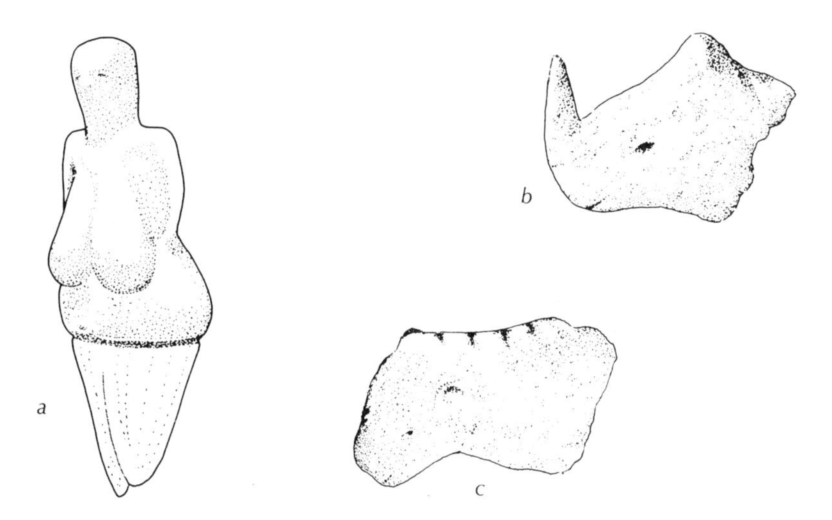

The "first" pottery. Figurines molded in a mixture of clay, fat, and bone ash and then baked hard; they were found in the remains of a mammoth-hunters' camp at Vestoniĉe, Czechoslovakia. At left is a "Venus" figurine (*a*); at right, a recognizable rhino head (*b*); and at center (*c*) what appears to be a feline of some kind. They pre-date man's first use of clay for pots by perhaps 10,000 years.

mammoth. Whereas the mammoth bones associated with Neanderthal and later campsites in the Ukraine had evidently been picked clean before they were collected for ballast, a mammoth-bone site at Predmost in Moravia tells a different story. The bone pile is estimated to contain the remains of 1,000 mammoths; it also gives every sign of being a kind of disassembly line — a butchering place. The mammoth skulls, laid open for extraction of the brain, are all at one part of the site, the hipbones are all at another. The mammoth hunters of Moravia even separated the tusks from the skulls and put them neatly to one side.

It staggers the imagination to think that all 1,000 mammoths were killed in the same hunt. Such a slaughter would require a hunting band far larger, composed of hunters far stupider, than we have any reason to believe existed in Upper Paleolithic times. Larger because 20 to 40 men, about as large a number of active adults and sub-adults as a hunting social structure can accommodate, are not likely to tackle even the most docile herd of mammoths that numbers more than 10 or 15 head. Stupider because, even if a regiment of hunters had somehow managed to assemble, common sense would call for a stop to the killing after, say, 20 to 40 of these walking storehouses of meat by the ton had been dispatched. To kill any more would be to lose the additional meat to putrefaction, even in winter.

At the same time, nobody who has killed a mammoth can do anything except butcher it on the spot. It is infinitely easier to carry off the meat than to try to move the cadaver. The only way I see to reconcile the difficulties presented by the Moravian bone pile is as follows. The terrain at the site must have been favorable for ambush and have also offered good grazing for mammoths. This being so, an annual winter hunt (involving a small number of hunters and the killing of no more than 8 or 10 mammoths on each occasion) would produce a bone pile of this size in a few generations. As to the sorting out of the mammoth parts, one can only suggest that perhaps the action became traditional in this part of Moravia.

I say a winter hunt because cold weather would retard spoilage long enough to permit the application of some kind of food-preservation technique such as is used by hunters today and was used even by our immediate farming ancestors. Meat can be sliced into thin strips and smoked over a fire. Once dried in this way, the strips will keep for months. As an alternative, the dried meat can be pounded into shreds and mixed with fat to form the kind of long-lasting iron ration we call pemmican. No one is likely to try to smoke a mammoth ham, but a reindeer ham might be manageable. Sausage? That seems unlikely, but I can imagine a kind of haggis without the oatmeal. In any event we should now turn to kinds of Paleolithic game more commonplace than mammoth.

Before dismissing the Moravian mammoth-bone site either as evidence of a curious annual event or as an enigma,

The Frozen Mammoth

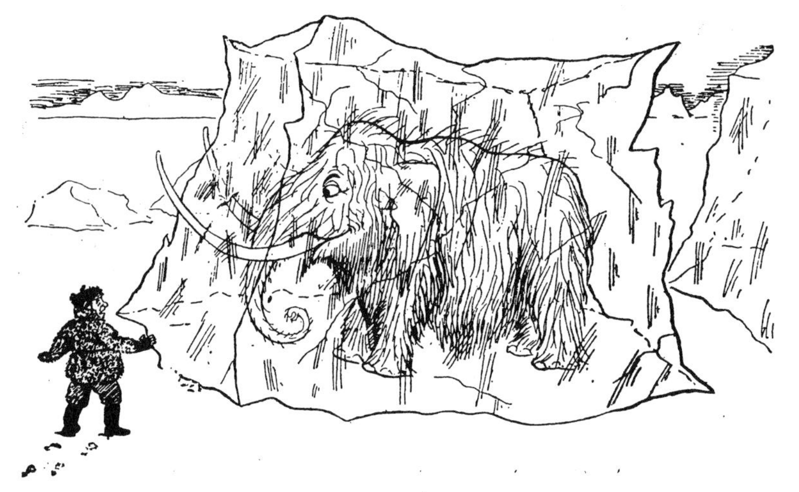

This Creature, though rare, is still found to the East
Of the Northern Siberian Zone.

It is known to the whole of that primitive group
That the carcass will furnish an excellent soup,
    Though the cooking it offers one drawback at least
        (Of a serious nature I own):

If the skin be *but punctured* before it is boiled,
Your confection is wholly and utterly spoiled.

And hence (on account of the size of the beast)
The dainty is nearly unknown.

That splendid versifier, Hilaire Belloc, created an imaginary problem for those who would dine on mammoth today. No less a problem faced the mammoth hunters of the Old Stone Age, even if it was different in kind. Unless they killed when the cold of winter would freeze the butchered meat, they lost most of it to bacterial spoilage.

however, it is worth returning to a parallel accumulation in France. There, north of Lyon, as noted earlier, a level plain ends abruptly in a steep cliff. Excavations near the foot of the cliff, at a site named Solutré, have unearthed the bones of 10,000 horses. There is scarcely any reasonable doubt about how this bone pile accumulated. Upper Paleolithic hunters — quite possibly the same skilled craftsmen who fashioned the lovely laurel-leaf flints of the Solutrean industry — regularly surprised groups of wild horses grazing on the plain and stampeded them over the cliff to sudden death. The hunters may have done this the year round, say once a week or once a month. In any event, whereas the pony-size horses of the Upper Paleolithic were very much smaller than mammoths, if they were slaughtered at the rate of 10 per week, the time needed to accumulate 10,000 cadavers at the foot of the cliff would have seen only one generation of Solutrean hunters succeeded by another generation. Thus precedent of a kind exists, both in the number of animals and in terms of a probably repetitive hunting practice, for the mammoth-killers of Predmost.

Back to game. We have already encountered mammoth and horse and have mentioned reindeer in passing. The patient study of the thousands of reindeer antlers found at Upper Paleolithic sites in western Europe has revealed something about the hunting practices of the time. Reindeer cast their antlers each year in the fall and grow a new set the next year. This makes it possible to deduce from the state of an antler just when during the annual cycle an animal was killed. The antlers found at the rock shelters in which the hunters took refuge during the winter are almost always cast ones. They were picked up as raw material for tools, rather than being left over from an early-autumn kill. Where, in the absence of rock shelters, the winter shelters were skin-covered tents — the discovery of postholes at the Ukraine sites attest to such shelters — the skins of these Upper Paleolithic structures were weighted down with cast antlers. With the arrival of spring, however, the hunters left their winter shelters and set up temporary camps in open hunting grounds. The antlers found at sites like these are new-grown; they came to camp still attached to a reindeer.

The bones of Old World bison — the aurochs — and of the species of wild cattle formally known as *Bos primogenius* are also present in the trash of Upper Paleolithic sites, bringing the inventory of big-game animals to a total of five. The hunters did not, however, pursue big game exclusively. The bones of geese, swans, willow grouse, and ptarmigan show they were fowlers of considerable skill. Now, ground-feeding land birds like grouse and ptarmigan can be snared, and a society that makes needles can produce twine for snares. But what about the water birds?

Were they stunned or crippled with throwing sticks? The implement is so simple that I have assumed its presence in men's hands, along with a digging stick and a cudgel, from the earliest time. Now, in order to knock down a water bird with a throwing stick, you hide in a blind built close to a feeding or resting area while your hunting companion comes up on the flock from the opposite side and spooks the birds into flight. If some of the birds pass over your blind still flying low, then you throw your stick and hope for the best. The earliest such throwing stick to survive comes from a peat bed in South Africa. Radicocarbon tests show that it is about 40,000 years old, an age that by European standards would place the implement in early Upper Paleolithic times. So we actually have a throwing stick and we know a practical way of using it with waterfowl. Was this done? I doubt it; it is more likely that the African implement was used to kill rabbits. I say this because there is a much better way to capture waterfowl. This is netting.

Go back to the blind and take another hunter. Carry with you a length of net — say fishnet — strung between two long sticks. When the spooker does his business and the startled birds fly toward the blind, you and your partner jump up at the last moment, stretching the net tight and holding it up at flight level. With luck you will net at least one bird and perhaps two or three.

Now, the earliest nets that have survived were not made until the end of the Würm glaciation, during the period of prehistory, mainly peculiar to Europe, called the Mesolithic. Therefore the presence of nets in the hands of Upper Paleolithic fowlers is only a guess. Nevertheless, the guess is given support by a line of reasoning that says: "Needles have eyes, eyes mean thread, thread means twine, twine means cordage, and cordage allows the weaving of nets." In any event, however they did it, these hunters killed and ate many kinds of birds.

They also caught and ate fish (nets again?). Amid the Upper Paleolithic trash are the backbones of trout and salmon. The salmon was also a popular subject in art, easily identified by its peculiar undershot jaw. Salmon were drawn in outline on bone; one, in the Dordogne, was painfully chiseled into the ceiling of a low limestone cave. Another bone carving of a fish is recognizably a pike, so we may assume that pike were also caught and eaten. Why is this, when we refuse a similar connection between a cat-decorated spear thrower and an appetite for cat flesh? Simple; an artist can grow familiar with the form of a cat without having killed or eaten any. Until he has the pike out of the water, however, he will not have seen enough of the fish to know its outlines.

Finally, of course, the hunters ate various plant foods. If only the science of pollen analysis had been known when these Upper Paleolithic sites were first excavated, we might now know quite a lot about what plants they ate. In any event, the digging-stick weights suggest a continuing interest in roots and tubers. Good-size cobblestones have also been found with a hollow carefully pecked into one side. Are these primitive mortars? If so, perhaps various seeds and nuts were pounded in them before being eaten.

I like to think, too, that they gathered berries and that perhaps a basketful or a skinful of berries, too long exposed to the summer sun, fermented spontaneously one day, thus acquainting Upper Paleolithic man with the delights of home brew thousands of years before grape and grain were willfully fermented. In this connection they may already have learned by dangerous trial and error, as did the historical hunting folk of Siberia, that certain mushrooms were agreeably intoxicating.

Whether or not the hunters of the Upper Paleolithic were getting high from time to time, they were certainly dressing better than man had ever done before. The evidence is present in their burials. For example, in the mid-1960s a 33,000-year-old grave was uncovered 100-odd miles northeast of Moscow. Its contents had been preserved both by burial and by cold. The original grave had been a shallow oval depression; the diggers first filled it with a layer of coals from a bonfire and then sprinkled the bed of ashes with red ocher. On this prepared surface they placed the fully-clad body of a 55-year-old man who was five feet, eight inches tall. He wore tailored fur trousers and a fur pullover shirt; both garments had been decorated with numerous rows of small mammoth-ivory beads. Around his neck were strung necklaces made of fox teeth, and on each of his wrists were a dozen thin bracelets made of mammoth ivory. Also buried with the hunter was a single stone pendant, a flint knife, and a large flint scraper; the two stone tools may indicate that he was a leatherworker in his leisure time.

TABLE THREE

PERIOD	NON-MUSCLE ENERGY AVAILABLE	MACHINES
Riss II through Würm II Middle Paleolithic (Up until 40,000 B.P.)	FIRE	*Lever class:* digging stick *Wedge class:* hand ax
Interstadial through Final Würm Upper Paleolithic (Up until 12,000 B.P.)	FIRE	*Lever class:* digging stick, spear thrower *Wedge class:* bone chisel

Here is a rich burial, preserved by chance. We have no hint of what our various Neanderthal dead may have been wearing at the time of burial. But whether it is the tailored garments that catch our eye, or the grave goods and the sprinkling of red ocher, we find in this grave further proof of the remarkably high material and spiritual quality of life in Upper Paleolithic times.

In summary, then, these hunters at the end of the Ice Age ate well and quite possibly were able to store provisions enough to assure eating well the year round. In all man's long history up to this point we have seen him capture little increments of spare time by doing things more and more efficiently. The Upper Paleolithic may be the first occasion when these random efficiencies combined to provide what we find commonplace today: substantial leisure time.

This moment, when the final retreat of the Würm ice sheet is ushering in a new era of prehistory, seems a good one for another summary of man's material progress. I will, as before, repeat part of the previous summary for the sake of continuity:

KNOWN MATERIAL CULTURE	NOT KNOWN BUT PROBABLE	EFFICIENCY 0.1 HP. x 12 hr. equals 50%
Mousterian flake-tool tradition (after *ca.* 100,000 B.P.). Firemaking kits. Use of man-made shelters (after *ca.* 60,000 B.P.). Wood lances. Ceremonial burials.	Dressed skins for clothing. Compound tools. Bone needles.	on evidence of shelter construction and time devoted to ritual, increase to 70%
Small tools made on fine blades. Bone needles, toggles; tailored clothing. Projectile points of bone & antler. Decorated artifacts. Portable & semi-permanent shelters. Lamps, fired clay. Ceremonial pressure-flaking. Ceremonial sculpture & painting.	Threads & cordage; snares & nets. Elaborate containers such as baskets. Practice of barter or trade.	on evidence of non-functional esthetic activities as an indication of increased leisure, increase to 75%

CHAPTER
6

New Energy Investments

THE FACE of Europe 10,000 years ago would seem strange to us today. The Scottish Highlands and most of Scandinavia were still buried in glacial ice, but men could have walked from the Atlantic coast of Ireland to the Ural mountains without wetting their feet. Scotland, Ireland, and England were joined to Europe by a broad plain that reached from County Cork to Brittany and kept the North Sea at bay from about the Firth of Forth to the northern extremity of Denmark. A great ice lake filled the basin of the Baltic Sea-to-be. This broad plain continued eastward to the Urals and, depending on fluctuations in climate, it was either pine forest or Arctic tundra. The woods clothing the higher land to the south also reflected the changes from cold climate to warm. In warm centuries the hardwoods advanced northward but they were replaced by pines during cold centuries.

This was a period of transition — from the Old Stone Age to the New Stone Age — known as the Mesolithic, or "Middle Stone," Age. The most important thing to remember about the Mesolithic is that it lasted for different lengths of time in different parts of the world. In the Old World it was shortest in the lands around the eastern end of the Mediterranean and onward to the east from there: the Persian Gulf and the Caspian Sea. Half a world away, in Southeast Asia, the Mesolithic seems to have been equally short. As to the New World, we have almost no idea whether the people of that hemisphere underwent any lengthy transition between being hunters and collectors and becoming farmers.

Even today a few societies in remote corners of the world continue to live as hunters and so, in that sense, the Mesolithic has still not drawn to an end. In Tierra del Fuego, certainly, it ended only a century or two ago, when the herders of sheep and the takers of guanaco hides advanced southward across Patagonia and exterminated the remnant Indian hunters there. (Until their destruction, these hunting bands tipped their arrows with tiny, carefully flaked silica points. They did not have to collect quartz or flint, however; their raw material was European bottle glass — empties, thrown overboard from ships rounding the Horn.)

A great deal of what we know about the Mesolithic has been learned from the many prehistoric sites in Europe, where, as far as its fringes are concerned, the period of transition ended only 5,000 years or so ago. This was about the same time the Bronze Age was reaching maturity in the Mediterranean. The reason why Europe is our main source of information is the same as before: more archaeologists have done more work there than anywhere else. Nonetheless, we are about to encounter a refreshing change. Some Mesolithic peoples also lived in other parts of the Old World.

Take, for example, the caves of Mount Carmel in Palestine, where the skeletons of various possible crossbreeds between Neanderthal and modern men were found. There, in what is now a dry and desolate valley, a team of English and American archaeologists excavated three caves between 1928 and 1934. One was known as the Cave of the Kid, another, the Cave of the Valley, and the third, simply, the Oven.

In these caves, underlying a scattering of Byzantine potsherds and classical odds and ends such as a Roman oil lamp and a figure of Aphrodite, but overlying the Neanderthal remains, the excavators found a stratum containing more than 17,000 stone implements, most of them flint blade tools of the miniature kind known as microliths. In among this imposing collection of tiny flint blades were at least 40 human burials — some individual, some in groups, and, in one instance, an adult with a child in its arms.

Several of the dead still wore circlets or entire caps on their heads, made from lengths of a tubular seashell, the genus *Dentalium*. Evidently these had been strung together on threads. Among the human bones were other kinds of beads and various pendants. In addition to microliths, the stone tools included choppers, picks, scrapers, and chisels, while artifacts made of bone included awls, pins, spearheads, and fragments of harpoons. The diggers also found limestone mortars, basalt pestles, and pieces of grooved sandstone; the grooves evidently were produced by rubbing the pointed bone tools against the gritty stone to sharpen them.

This was a rich find but, except for one further thing, it might not have captured the imagination of many prehistorians other than the excavators themselves. This further thing was located among the fragments of bone: thirteen narrow bits and pieces of bone, the largest of them some six inches long, marked on one edge by neat wedge-shaped grooves. Now, just such grooves were gouged out of narrow bone points in late Paleolithic and Mesolithic times in Europe so that tiny flints could be inserted and cemented in place, perhaps with pine pitch. The microliths provided such a "compound" projectile point with its cutting edge.

What was the reason for the grooves in the Mount Carmel bones? One bit of grooved bone the diggers found had two microlith blades still fixed in place. The investigators sorted over their thousands of tiny blades, setting aside any that could have fitted the bone grooves. They found that the edges of many of these blades showed a characteristic shiny polish. It is a polish produced by wear: the friction of cutting through the surface layer of silica on the stems of grasses.

The foremost example of silica-coated grass is the largest of all: bamboo. If you have ever splintered a bamboo fishpole you know that the edges of the splinters are razor-sharp. The surface layer of silica has been exposed as a cutting edge. Break it cleanly enough and you can shave with it if you want to. You may well ask why the excavators recognized this peculiar kind of silica polish on the edges of the Mount Carmel microliths. The answer is that they were familiar with such polishes from previous work at sites of the following period, the Neolithic, when flint-edged sickles were commonplace.

These lengths of grooved bones, then, could have been sickles. Today, to speak of a sickle, a tool for reaping grain, implies that the user of the sickle is one who plants a crop of grain and harvests it: in short, a farmer. In the late 1920s and the early 1930s, however, few were ready to believe that a valley population in Palestine whose level of development was merely Mesolithic could have already taken one of the key steps in man's capture of energy. How could mere cave dwellers have initiated the Neolithic Revolution?

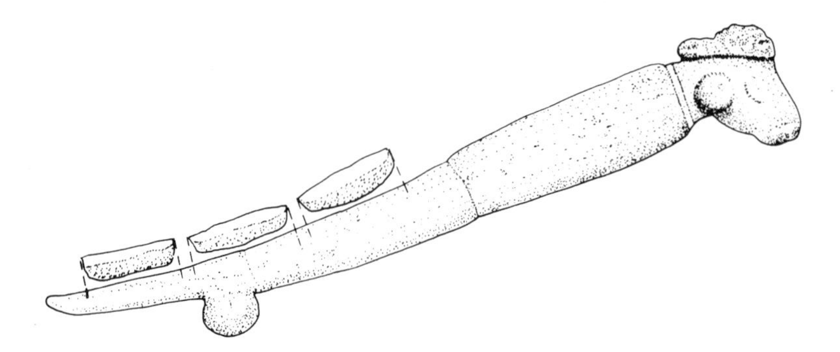

Natufian sickle was made by setting sharp flint microliths into a grooved bone handle, probably cementing them in place with something sticky like pine pitch. Silica polish on the edges of the microliths proves that they were used to cut grasses of some kind. Seed-bearing wild grasses? Not likely; wild grasses, when ripe, "shatter."

An explanation was quickly invented that avoided so revolutionary a conclusion. The Natufians, as these sickle-wielders were named by their discoverers, had probably used their harvesting implements to reap certain seed-bearing wild grasses: this would have added a vegetable supplement to their diet of wild game and fish. The presence of mortars and pestles — artifacts particularly suited to the task of pounding plant produce into edible form — strengthened such a conclusion. This was a comfort because, if the Natufians had merely gathered wild grain, the start of the Neolithic Revolution was still safely preserved as the accomplishment of other peoples at some time closer to the present.

Unfortunately for comfort of this kind, the scholars who put the Natufians back on the shelf as simple Mesolithic hunter/gatherers failed to consult either the botanists (who really know about subjects such as wild grain) or the ethnologists (who know very well how the "uncivilized" as well as civilized people of today behave). The botanists could have told them that the seeds of wild grasses, when ripe, are scattered haphazardly. This natural reseeding process is known to botanists as "shattering."

The wild grasses of Palestine, among them emmer, a then wild form of wheat, all shatter when ripe. The grasses that have come to be man's fundamental grain crops — wheat, maize, rice, barley, rye, millet, and sorghum — have all undergone a mutation that has made them non-shatterable. Such a mutation is, of course, suicidal unless the plant has become domesticated — that is, planted, harvested, and replanted by man. Having lost the power to broadcast its seed, the no-longer-wild grass cannot survive without man's help.

If the investigators at Mount Carmel had consulted the ethnologists, these students of primitive peoples could have told them how "shattering" grain is harvested today. For instance, the Winnebago Indians of Lake Superior, who go in canoes to harvest the wild rice that grows along the shallows of the lake shore, have no use for sickles. Should they try to cut the ripe stems as wheat harvesters do, bundle them up and carry them off to a threshing floor, most of the shattering seeds would be lost. The Winnebago avoid this kind of loss. They harvest the wild rice by threshing on the spot, bending the uncut stems over

The Indians of Lake Superior still take advantage of the way a ripe wild grass "shatters." In their canoes they go into the offshore stands of wild rice, bend the ripe ears inside the canoe, and tap them until the grains of rice fall off the stalks.

the gunwale and beating the stem heads until most of the shattering grains have fallen into the canoe.

All this being so, what did the Natufians do with their primitive sickles? If they used them to harvest ripe wild grain, then the grain they cut down must have already mutated into a non-shattering variety or there would be no harvest. But a non-shattering mutation, because it cannot reseed itself, must be a cultivated rather than a wild grain. If we follow this line of argument, a Natufian sickle is proof of a cultivated grain crop and this makes these Mesolithic Mount Carmel people early possessors of one of the skills — growing crops — that characterize the Neolithic Revolution. Perhaps this is so; the thought is not so radical today as it seemed half a century ago.

For those who wish to keep the Neolithic Revolution safely closer to the present, still other explanations are possible. Although there is not a scrap of evidence to support the conjecture, the Natufians may have wanted to make containers. Like the peoples of the Upper Paleolithic period who preceded them, these folk of Mount Carmel were familiar with cordage. For example, the threaded shell circlets and the beads found in their burials attest to their use of thread. Like their predecessors, living in the same place for a considerable time, they must have felt the need for storing things such as surplus food in a convenient fashion. For that matter, containers are also useful when shifting from, say, winter quarters to summer camps. I have already suggested, in spite of lack of evidence, that skins and baskets were used as containers long before Mesolithic times. Why not imagine baskets at Mount Carmel?

One reason why the Natufians may have needed sickles: to cut grass suitable for making storage baskets. This requires a lot of grass and even more patience, but one can start off small (*three steps at top, l. to r.*) and end up with a large basket (*bottom*).

The simplest kind of basket — still made in school craft classes today — is put together as follows. You hold in your hand a small number of grass (or straw) stems so that they form a sausage-like cylindrical bundle. Next you begin to wrap a cord around the bundle. About halfway up the cylinder, which will now hold its shape thanks to the wrapping, you stop and collect another handful of stems. You stuff the ends of the second handful into the loose ends of your half-wrapped bundle and resume the wrapping process. Although each of the stems in the bundle may be short (baskets like these have even been made out of pine needles), you can produce a wrapped sausage that is several yards long.

If you now coil the sausage on itself in a spiral, and tie each full circle to the one underneath it, you can make baskets that are cylindrical, bottle-shaped, dumbbell-shaped, or even flat. Such baskets are not watertight but they make excellent containers for solids. Now, if we suppose that the Natufians needed baskets (and also suppose they made them in this simple manner), then we have found a non-grain-harvesting use for their sickles. These compound tools of bone and flint would have served to harvest an abundant supply of grass stems for the basketmakers. Perhaps this was so.

An even simpler explanation of the Natufian sickles exists. Imagine that these Mesolithic people did indeed harvest wild grasses with their sharp-edged implements but avoided the shattering of the ripened heads of the grain by cutting the stalks before the heads were ripe. This would eliminate the loss due to shattering. Perhaps this, too, was so, although such a practice raises a further question. Is not such a thoughtful and systematic practice of collecting a wild foodstuff almost the equivalent of farming? If your answer is "yes," then the Natufians must at least be honored for standing on the threshold of a new age.

I think they should be honored even more. Anyone may collect green heads of grain and dry them out. If, however, someone goes to the considerable trouble of making stone mortars and pestles, it seems fair to assume that the intention is to pound the harvested grain into some kind of edible product. Ripe seed is a better foodstuff than dessicated green seed.

Sickles in the hands of the Natufians some 10,000 years ago, then, are probably evidence that someone, somewhere between the Mediterranean and the Caspian Sea, had already begun to tend mutated grasses and had thus accomplished one of the two major advances of the Neolithic Revolution. Still, it would be comforting if pollen analyses or even the discovery of a few grains of mutated barley, oats, or wheat were to turn this probability into fact.

The other major Neolithic advance, as we will see, was animal domestication. Again, we find the event foreshadowed in Mesolithic times. Man's first domestic animal was the dog, and the first evidence that men and dogs had begun to associate appears at this stage of prehistory. One of the animal remains at a famous Mesolithic site in England — Star Carr in Yorkshire — was the skull of a dog. Its teeth are relatively large in comparison to the size of the jaw and thus appear "crowded." This kind of crowding, in the view of Barbara Lawrence of the Museum of Comparative Zoology at Harvard, demonstrates that the Star Carr dog was domesticated rather than wild. Star Carr has been dated by the carbon-14 method; it is about 9,500 years old.

This dog (which, by the way, was small) may be the oldest known domestic dog in Europe, but it is not the oldest in the Old World. That title is reserved for the animal whose fossil jaw was found along with evidence of human occupation in a cave in Iraq in the 1950s. The dog had lived with the men who took shelter in Palegawra Cave some 14,000 years ago. So far, the Palegawra dog is the earliest domestic animal known.

Is there also an oldest-known domestic dog in the New World? Yes, two of them. The honor is reserved for dogs of different sizes whose remains were found at a romantically named site in Idaho, Jaguar Cave. During the summers of 1962 and 1963, archaeologists visited this natural shelter in Birch Creek Valley, on the west slope of the Beaverhead Mountains in

Lemhi County. They found the jawbones of the smaller of the two dogs buried near a hearth; the carbon-14 age of charcoal from the hearth proved to be some 10,400 years. The remains of the second dog, about the size of a small wolf, came from the same level at Jaguar Cave; it is therefore probably about the same age.

Now, variation in size is one of the consequences of domestication in animals. Thus the discovery that domesticated dogs of different sizes existed in the New World more than 10,000 years ago has immediate implications. First, both the practice of domestication and, in particular, man's association with dogs must be substantially older than the age of the Jaguar Cave charcoal. Second, the two dogs had not been brought into domestication from any canine stock native to the New World. This means that the Jaguar Cave pair must have been descended from domesticated dogs brought to the New World still earlier.

It would be asking too much of coincidence to expect that the Idaho pair were the very first two dogs to arrive from the Old World. Even if this highly improbable assumption were true, however, mankind's first association with dogs evidently extended farther into the past than the Iraq find of 14,000-odd years ago would suggest. Why? Because we must allow a reasonable amount of time for men and dogs to have lived together before two distinctly different breeds of domestic dog would evolve. We must also allow still another span of time (which might in part have run concurrently) for men with two breeds of dog to emigrate from the Old World to the New. Fourteen thousand years ago the men who lived in the Palegawra Cave had probably reached the Mesolithic level of culture. Outside this broad zone between the eastern Mediterranean and the Caspian Sea, however, most of mankind was still making do in the Paleolithic manner. If we admit that men and dogs may have teamed up much earlier than the date of the Palegawra site, then we are in fact admitting the possibility that Paleolithic man was the first to domesticate an animal.

How did man and dog first meet? In different parts of the world today dogs are variously cherished as house pets, left to forage for themselves, or fattened for the table. But many dogs today are also carefully bred and trained as companions to men in hunting and herding. Therefore, while we know nothing at all about man Number 1 and dog Number 1, the dog's skill as a tracker and attacker in the hunting field may have been what first made men want to win the help of dogs.

A well-known biological principle is labeled "symbiosis." This is a scholarly way of saying "living together" (*sym* is the "together" part of the word and *bio* is the "living" part). Symbiotic relations exist in a nearly endless variety of ways between various animals, between animals and plants, and between various plants. Let me give two examples. First, the lichens you find growing on tree stumps in the woods are not individual plants. Instead they are a complex partnership between two quite different plants: a fungus and an alga. You can, if you wish, consider a lichen a kind of corporation with mushroom directors and seaweed stockholders. A second example is provided by a kind of stinging ant that has taken to living in hollow parts of a species of acacia bush. If a passing goat tries to nibble away at such a bush, the disturbed ants swarm out and bite the goat. In payment for this service, so to speak, the ants feed on the nectar, pollen, and berries produced by the acacia.

Just how any of these and thousands of similar relations were first established is an unanswerable question. One can only say, in the most general way, that this kind of living together appears to be somewhat more efficient than living apart. Accepting this as true, it follows that the animals or plants that develop such partnerships have an adaptive advantage — that is, a better prospect for survival — over those who do not.

This is, of course, a wishy-washy answer. Lots of seaweeds and mushrooms flourish independently, lots of bushes survive without attendant ants, and lots of ants make do without sheltering in bushes. Similarly, many canine species — and many hunters — live independent lives. Nonetheless we know that men and dogs did, at some time in the past, enter into just such a symbiotic relation. We also know, from the evidence at Star Carr, at Palegawra, and at Jaguar Cave, that the relation is substantially more than 10,000 years old. Because the men of those times were hunters first and foremost, a hunting partnership seems to be the likeliest first relation between man and dog.

Both the keen-nosed dog, wolfing down (if you'll pardon the expression) the scraps from the hunter's kill, and the hunter, seeking out prey in the forest thickets, would have had something to gain by such an association. It may be, in the extreme example, that the symbiosis had a genuine adaptive advantage. When the game was thin, hunters helped by dogs may have kept eating while hunters without dogs starved. What seems more likely to me is that hunters with dogs simply fared so much better that the dogless hunters, learning from the example, went off and found dogs of their own.

Two other man-and-dog items demand mention. First, not only is the dog man's earliest domestic animal, but it is also the first animal whose muscle power man was able to harness. Thus at last, in a Mesolithic context, man had gained control of

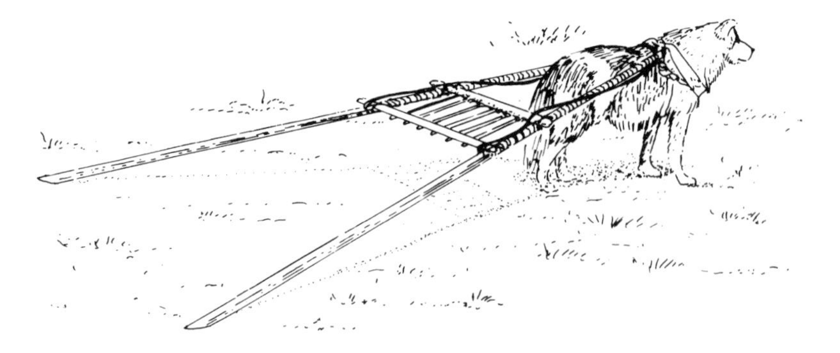

The dog was man's first "domesticated" animal. The earliest evidence for this in the Old World is dated about 12, 000 B.C. and in the New World about 11,000 B.C. Surely at first a hunter's helper, the dog soon enough became a draft animal. At the left is shown a Plains Indian travois, or pole-sled, that allowed a dog to draw a useful load in the wheelless New World. At the right are some cart-pulling dogs of New York City in the 19th century; the scene is American but the practice was introduced from Europe where, during the Middle Ages, miners even loaded dogs with sacks of ore.

an energy source other than the now familiar trio: fire, his own muscle power, and the assistance of the simple machines. Considering that man's own muscle-power rating is only one tenth of one horsepower, a dog's muscle-power rating is not very impressive. Nevertheless, dogs were still being harnessed to small carts here and there around the world until the end of the 19th century. In the wheelless pre-Columbian New World, the Indians of the North American plains harnessed their dogs to the baggage skids known as *travois.* If the great myth of India, the Rig Veda, may be trusted as a source, early farmers of the subcontinent even harnessed dogs to plows. "You two plow the barley with the wolf," the line reads; we need hardly believe the writer meant a wild wolf.

The other item comes from the world of modern medicine. If you think about swine flu for a minute you will realize that men share a number of diseases in common with their domestic animals. The classic example, indeed, is Edward Jenner's demonstration that cowpox was a mild form of the virulent killer of

men: smallpox. A less familiar example is distemper in dogs; the disease is a close kin to measles in humans. In the absence of immunity, both forms of the disease are likely to be fatal, as the Indians of the Americas learned to their sorrow when the Europeans reached the New World.

In any event, medical scholars have tabulated the diseases shared by man and various animals. If we make the assumption that, in general, the more ancient the association, the more diseases there will be in common, then the scholars' tabulation represents evidence of a kind for the sequence of man's successive animal domestications or similar close associations. The tabulation in the early 1960s was as follows:

Men and dogs —	65 diseases
Men and cattle —	50 diseases
Men and sheep and goats —	46 diseases
Men and pigs —	42 diseases
Men and horses —	35 diseases
Men and rats and mice —	32 diseases
Men and poultry —	26 diseases

Fowl actually came into domestication quite early, so that the low rank given to poultry in this tabulation probably represents an essential biological difference between birds and mammals so far as harboring disease is concerned, rather than a measure of the length of man's association with chickens, ducks, geese, turkeys, and similar domestic fowl. By the same token, few mice and scarcely any rats are domesticated except in the laboratory. Instead, these rodents are commensal with man. (The word is another scholars' way of defining a kind of symbiosis: *co* equals "together," as in "coed," and *mensa* equals "table," so that the intended sense of the word is "eating at the same banquet board.") Rats, mice, and man began to eat at the same table as soon as man began to store foodstuffs, grain in particular. Being commensals does not, however, demand physical contact. Thus, whereas rats in particular have been the carriers of diseases that are particularly virulent to men, they rank lower than horses (which men ride, feed, and groom) or pigs (which may be hand-fed and can even be house-guests) or sheep, goats, and cattle that are variously sheared, hand-fed, herded, milked, and treated like family.

But notice who heads the list: Arf! As a humble commensal with a cat, I am only sorry that the medical scholars did not see fit to include those arrogant semi-domestics in their disease table. To digress briefly, logic suggests that cats first found putting up with men worthwhile when men's storage of grain began to attract rodents in numbers far greater than are found in the wild. To be free to plunder a virtual zoo would have been an attractive prospect. This hypothesis would place the beginning of cat-and-man relations early in Neolithic times (or possibly late in Mesolithic times among those communities that may have collected enough wild grain to be worth storing). Just how good a job the cats did we can only guess. In Neolithic Iran around 2500 B.C., at least, their performance was not so tiptop as to smother man's invention of the mousetrap.

To summarize, we have the possible beginning of agricultural practices in Palestine some 10,000 years ago, which implies even earlier experiments with wild grains elsewhere in western Asia. We also have certain evidence of man's domestication of the dog in adjacent Iraq some 4,000 years earlier (with the similar implication that dogs were first domesticated still earlier). These are two notable Mesolithic milestones, foreshadowing the Neolithic Revolution-to-come. What else did Mesolithic men accomplish? Let's start with how they hunted.

If western Asia was a paradise of sorts in Mesolithic times, the dank woodlands fronting on the now-drowned North Sea coast must have seemed the kind of cold and awesome Hell that is so vividly described in Scandinavian myth. The folk of Mesolithic Europe hunted and camped in the inland forest vastness, as the remains of forest animals attest, but they took as much or more of their living from the sea, the marshlands, and the lakes and streams of more open coastal areas. The shells of the oysters, clams, and mussels they gathered between the tides were discarded a short distance inland in great mounds that archaeologists call "kitchen middens." As those who have ever stood downwind from a heap of discarded scallop shells will readily appreciate, it speaks much for the fortitude — or the limited sense of smell — of these Mesolithic peoples that they actually camped on the middens. Nevertheless, the remains of hearth fires, the bones of other animals, and these hunter-fishers' tools and weapons are found buried deep in the shell mountains.

Along the coast the hunter-fishers caught cod and flounder, presumably using boats of some kind. From the lakes and streams they took pike, eel, roach, garfish, perch, and stickleback. Some of the fish they caught on lines; the hooks have survived. Some they netted; net floats, net weights, and even

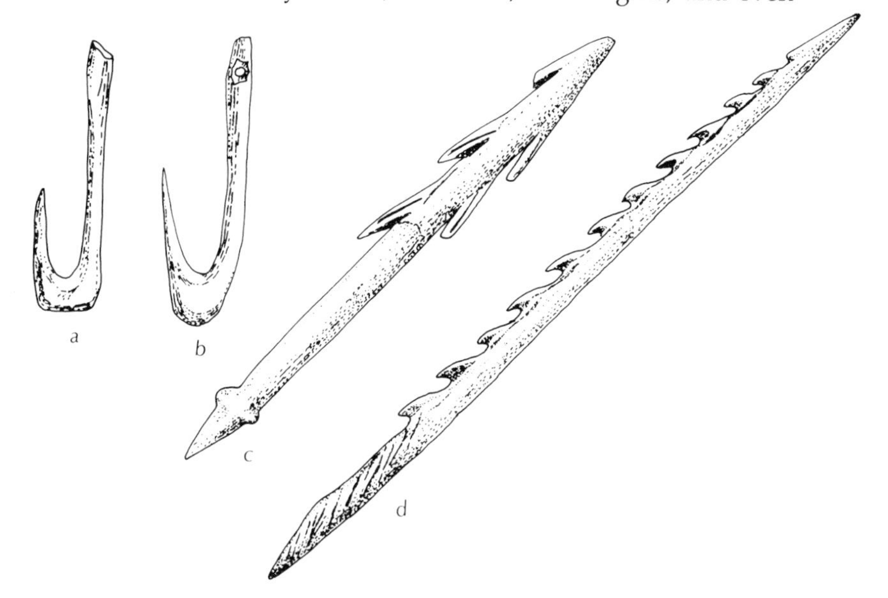

Fisherman's tools: although not barbed, as most hooks are now, these two (*a, b*) are recognizable for what they are although they were made nine thousand years ago. The harpoon heads (*c, d*) are also Mesolithic implements, probably for use in streams.

some bits of netting have survived. Some they speared; bident and trident spear points have been found in abundance.

First rank among the animal bones, however, is reserved for waterfowl. They killed shoveler, pintail, scaup, goldeneye, scoter, merganser, and tufted duck; gray goose, brant, whooper swan, and mute swan; grebe, herring gull, guillemots and other auks, heron, bittern, cormorant, and pelican. In addition to waterfowl they killed such swift raptors as the kite, osprey, and white-tailed eagle (the latter, perhaps, because they wanted to fletch their arrows with its handsome barred feathers).

Yes, these Mesolithic inhabitants of Europe had added a new weapon — the bow and arrow — to their arsenal. There is

A major Mesolithic advance: the bow and arrow. In the thick forests of Europe using a spear thrower was impractical, which may have hastened the development of the bow as a hunter's weapon. This vivid scene of hunters armed with bows is from Spain.

an abundance of evidence for this. For example, a Russian Mesolithic grave — on Deer Island in Lake Onega, northeast of Leningrad — proved to contain a cache of arrows. The arrow shafts were made partly of bone and partly of wood; the projectile point was hafted to the wood part of the shaft. All the points were made of flint flaked in a manner typical of a flint industry that flourished in western Russia and adjacent Poland between 8500 and 7800 B.C. From Denmark come more wooden arrows with their projectile points lashed in place and, better still, two bows, handsomely shaped out of ash wood.

The development of the bow, a device that seems quite trivial to us today, is actually a major milestone in man's control of energy. Why? Because the bow is surely man's second — and perhaps his first — "storage battery." The concept underlying the bow is that of the spring: a device capable of storing energy that is invested in it little by little, as when you wind a watch. Later the stored energy can be released, either little by little, as with a watchspring, or all at once, as with a bow.

As the archer steadily pulls back his bowstring, making the springy bowstave bend, he is storing energy in successive increments. Take as an example a bow with a 50-pound pull that can fire a 31-inch arrow. At full pull, the archer has usually brought the bowstring 24 inches back from its neutral position. Multiply two feet by 50 pounds: the bow-and-arrow system is now storing 100 foot-pounds of potential energy. What then?

Perhaps this is the place to remind you again that energy is the capacity to do work, but that power is something different: power is the application of that capacity over a specific period of time. To repeat the basic example, one horsepower is the power that is generated when 550 foot-pounds of potential energy are transformed into kinetic energy in a single second. To repeat further, the horsepower rating given to man's muscular capacities is 0.1 horse, or one tenth of one horsepower. This is to say that when a man and a horse are set to work over any period of time — one second, one hour, or a full day — the horse will accomplish ten times the work of man.

Early bow, ultimate bow. The flat, only slightly shaped wood bow (*top*) is from the Mesolithic of Denmark; the clutch of arrows beneath it, with flint arrowheads and wood and bone shafts, are from the early Mesolithic of western Russia. The ultimate bow (*bottom left*) is the weapon of a Tartar horseman of the 14th century, made of laminated wood and horn. It seems misshapen until strung (*bottom right*).

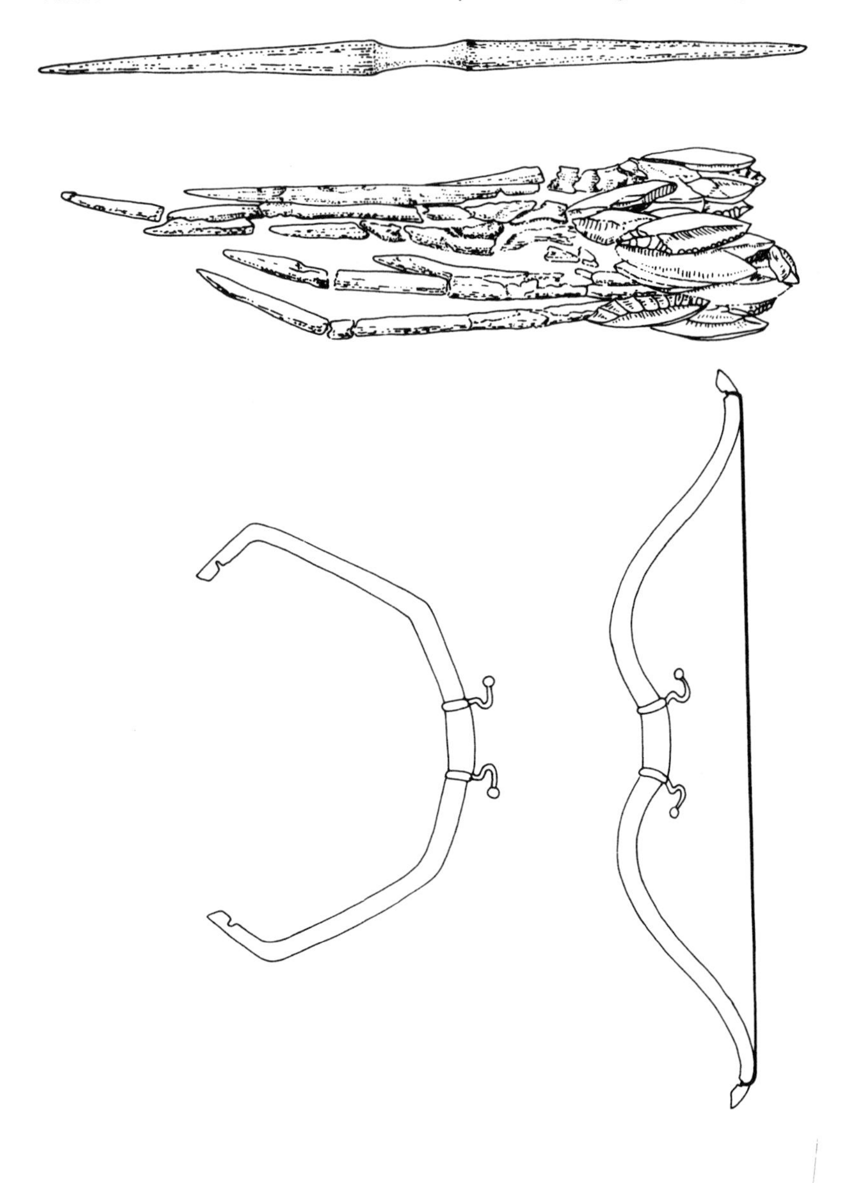

Now consider the archer. Say that he takes 20 seconds to draw his bowstring back and that, during those 20 seconds, he has invested one tenth of his muscular strength. In this, as in all energy investments, one never gets something for nothing. Energy may be transformed, but it can neither be created nor destroyed. So the archer's investment is equivalent to one hundredth the energy a horse could have invested in the same length of time. (How incongruous! But never mind, think of Sagittarius.) Cancel this sum out on both sides of a balance between 20 seconds and 0.01 horse and you find that the archer's investment is one fifth of a horsepower. If it takes one second for the bowstring to snap back from full-drawn to neutral, this fifth of a horsepower is very nearly the power the archer delivers to the arrow.

Does this mean the archer has involved himself in a meaningless investment? Not at all. For one thing, he has exchanged energy slowly accumulated for energy swiftly applied. For another, a factor that I find easiest to think about as the "seesaw effect" has been invoked.

Imagine that you have tilted a weightless and frictionless seesaw until one end touches the ground. Now place a pebble on this down end. Next you could drop a heavy rock on the other end of the seesaw. The down end would rise with a snap and flip the pebble up high overhead. How fast — and thus, how far — the pebble travels would be proportional to the difference between the weight of the falling rock and the weight of the resting pebble. By analogy, in the archer's case, his release of the bowstring represents the falling rock and the lightweight projectile that flies from his bow represents the far-flung pebble.

Again, to be sure, nobody gets something for nothing. Nevertheless, the greater the velocity of the arrow, the greater will be its impact when it strikes its target. As an example, the shock effect of a one-ounce projectile traveling at 1,600 feet per second is equal to the shock effect of a one-pound projectile traveling at 100 feet per second.

Of course arrows do not travel at such high speeds. Even

with modern souped-up bows and extra-light arrows a "muzzle" velocity in excess of 200 feet per second is remarkable. A Mesolithic archer would have been lucky to loose an arrow at half that speed. All the same, the seesaw effect is recognized by today's archers. They know from experience that a light arrow travels faster than a heavy one and that a small difference in weight translates into substantially greater "muzzle" velocity for the lighter shaft.

To summarize all this, the bow and arrow is a fine weapon and the spring is one very convenient means of storing invested energy. What did this mean to the Mesolithic archer in day-to-day terms? Let us see. For one thing, among the Mesolithic arrows that have been unearthed are wood shafts with carefully carved blunt points, twice or three times the diameter of the shaft itself. By analogy with known archery practices, the main use for a blunt arrow is hunting fur-bearing animals. If the wallop of a blunt arrow will knock a squirrel out of a tree or stun a rabbit, why put a hole through its skin? The Mesolithic hunters also tipped arrows with the tiny flint points they made by striking thin blades from a core of flint and then breaking the blades into useful shapes. There is ample evidence that such small projectile points were effective. For example, the skeleton of a bison was found buried in a peat bog in Denmark. Small flint arrowheads were found stuck in two of the beast's ribs, and others, evidently from the tips of arrows that had not struck bone, were present inside the rib cage.

Evidence also exists that not every target of the Mesolithic archer was an animal. From a kitchen midden in France comes the skeleton of a young man with just such a flint arrowhead driven for a third of its length into his spine.

The bow was not used for hunting alone. Here in Mesolithic times we find another use for a length of wood that can be drawn into a curve by means of a length of cord. When combined with a "bit" — in this instance a pointed flint hafted to a stick — this kind of bow is a drilling machine. In principle it is the same as the fire drill we may have already encountered in Paleolithic times, but the bow drill is far more sophisticated.

We know about Mesolithic bow drills because one hunting group in those distant days liked to decorate bits of bone and antler in a particular way. I'll call this group the "Big Boggers" because their artifacts were first found at a site near Mullerop, in Denmark. The site was named Maglemose, which is the Danish for "big bog." The Big Boggers had a taste for abstract decoration: geometric designs executed in lines and dots. Some of the fine dots were simply incised with a sharp, hand-held flint point. A number of other dots, however, were pits cut into the surface of the bone and antler with such uniformity of shape that they could only have been produced by drilling. One of the Danish investigators who helped excavate the Big Bog site, Sophus Müller, discovered this when he examined the decorative dots

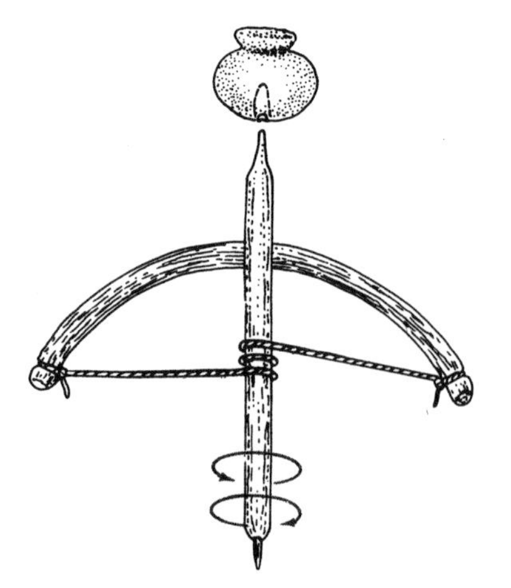

Halfway to rotary motion: the bow drill. Which came first, the bow or the bow drill? No one knows, but both were in use by Mesolithic times. With a drill possibly much like this one a marshman of what was to be Denmark decorated a length of antler with a neatly drilled row of holes. The archaeologist who found the decorated antler decided that the bit had been made of bone. Can you guess why?

on a bit of antler retrieved from the bog. He found that the successive pits increased in diameter from a 20th of a millimeter at the start of a row to a quarter of a millimeter at the finish. From this he concluded that a bow drill had been used and that its bit had become more and more blunted as the work progressed. Müller further concluded that the bit had probably been made of bone because a bit made of pressure-sensitive flint would have worn down much more unevenly. Müller's is a particularly elegant example of how much of men's past activities the archaeologist can reconstruct on the basis of tiny scraps of surviving evidence.

Invention of the bow drill, although it seems simplicity itself once the device is explained, was an awesome achievement. It led mankind halfway down the path to continuous rotary motion. The development of continuous rotary motion, in turn, is the fundamental accomplishment that underlies the generation of most kinds of power. Here, in Mesolithic times, these broader implications may not have stood out. Nonetheless, indirect evidence suggests that the inventors of the bow drill soon put it to a number of non-decorative uses.

If fowling inspired the bow, did fishing inspire the adz? Once again an accident of geography has helped provide an answer. Many of the places where the Mesolithic inhabitants of Europe lived were boggy because, as the last of the glacial ice of Paleolithic times turned back into water, the sealevel was slowly rising. Indeed, much of the ground that was still dry land then has since been turned into bog or drowned altogether. One happy aspect of this slow immersion is that the remains of large wood artifacts have been preserved by accidental burial in peat or mud.

One of the largest of these surviving Mesolithic artifacts is the remains of a 10-foot-long dugout canoe, unearthed from a layer of peat that lay buried under 10 feet of silty soil — once plain wet mud — near Perth, in Scotland. Neither bow nor stern of the canoe had survived in good enough condition to give the excavators much notion of what their shape had been. The wood of the main body, however, was recognizably Scottish fir. At its widest point the canoe measured three feet from side to side. The presence of charred wood on the inside surface suggests that the digging out of the dugout, the process that transformed a fir tree trunk into a means of traveling by water, had been assisted by fire-hollowing.

Until almost yesterday many "uncivilized" fishermen made logs into dugouts by means of fire-hollowing; the method greatly reduces the physical effort required to remove most of the wood from the inside of a cylindrical log. After the canoe-makers have found and roughly shaped a suitable log they build a series of small fires along its top. The fires are allowed to burn down, are fed back to life, and allowed to burn down again and again until almost all of the top of the log is merely charred wood. The fires are then swept off the log, the smoldering wood is doused with water, and the damp char is chipped away. This process is repeated until the log has been hollowed out.

The length of Scottish pine was surely dug out in a similar manner. Fire-hollowing, however, is not the main point here. What did the Mesolithic shore folk use to chip away the char? This could have been done with the naturally sharp edges of large sea shells; sea-shell adzes have been widely used for woodworking by the "uncivilized" islanders of the Pacific Ocean. Adzes made of stone, however, are much more durable, and, because of their greater weight, more efficient.

In the early days of archaeology the distinction between the Old and the New Stone Age was based on a technological change. The big flint choppers and hand axes of Paleolithic times were made by striking flakes from a core until the core assumed the desired shape. Most of the axes and adzes of the following Neolithic period, regardless of how they had been roughed out, were brought to their final shape by grinding.

But when did the practice of grinding start? As you might by now expect, seeing how many things happened earlier than was once supposed, the first evidence of shaping stone tools by

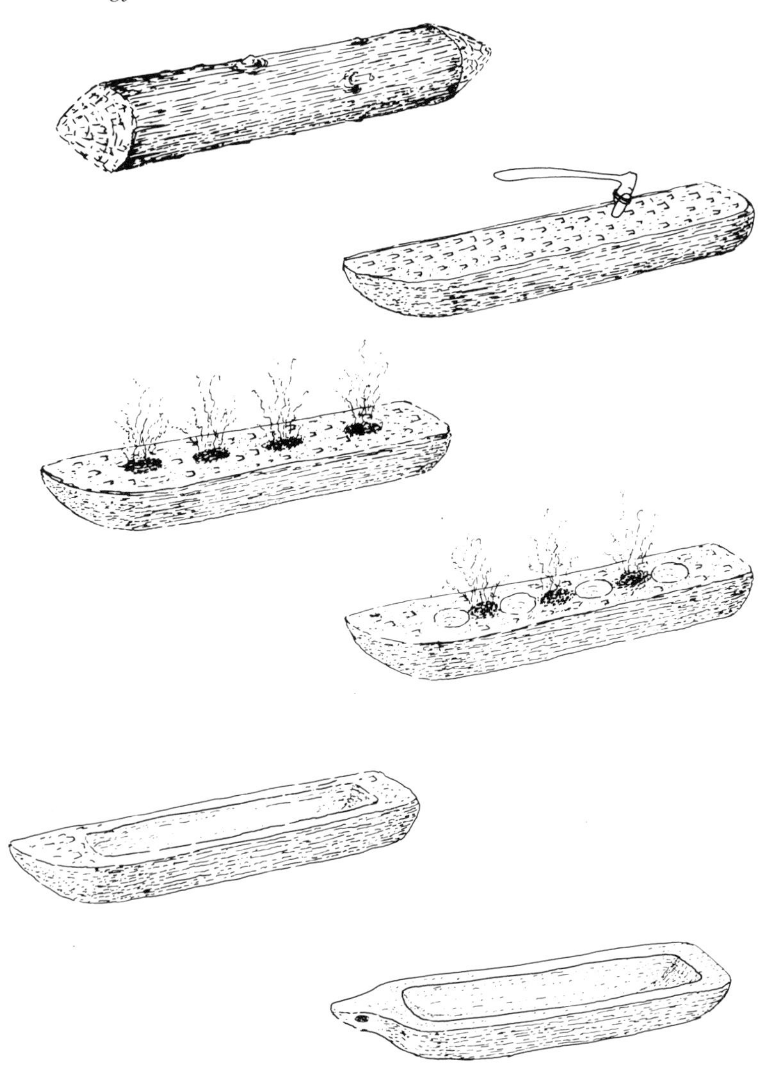

How to dig out a dugout. Having chopped down a suitable tree and tapered the log at each end, the proto-carpenter of Mesolithic times used an adz to cut the top of the log level. After that, fire will do most of the work. After each in the series of small fires burns itself out (or is quenched), the adz is used again to remove the charred wood. Burn and chop, burn and chop. In due course, a canoe.

grinding is found in this period of transition between Old Stone and New Stone, the Mesolithic. The tools Mesolithic men made in this fashion were still shaped mainly by striking waste flakes from a core, but the final cutting edge was made sharp by grinding. Axes and adzes of this kind began to appear in Europe around 7000 B.C. The Perth dugout was made at some time between 6800 and 5000 B.C.

At first these new tools were hafted in a complicated way. A length of antler was made into a "sleeve" by digging (or drilling) a hole into the wider of the two ends. The butt of the stone head was then fitted snugly into the prepared hole. The combination of antler sleeve and stone cutting edge could then be hafted in a split wooden handle and secured in place with lashings. Sometimes, instead, a hole was drilled all the way through the antler haft, at a right angle to the sleeve hole; the wooden handle was then inserted through the hole in the antler .

Up to now I have cheerfully imagined that even our most remote ancestors used wooden artifacts. We have seen later proof of this in the form of the Clacton-on-Sea yew shaft and in the woodworking polish on the edges of Acheulean stone tools of the same era. Later still, although very few wooden objects of the Paleolithic period have survived, we have not hesitated to assume that the spears that were flung with extra velocity in the Upper Paleolithic, thanks to the use of spear throwers, had shafts of wood. Nevertheless, it is only here in Mesolithic times that the evidence for woodworking on a large scale becomes abundant. Indeed, if we ignore the lack of saws, it would almost be possible to say that the Mesolithic witnessed the birth of carpentry. And all this came about with the help of these new ground-edge tools.

What did Mesolithic hunters make out of wood besides dugout canoes? We know they made bows and arrows, although the manufacture of these objects does not require tools any more advanced than those of Paleolithic times. What seems truly unique is the appearance at this time of objects made from planks of wood: staves for skin-covered (or bark-covered) boats, pad-

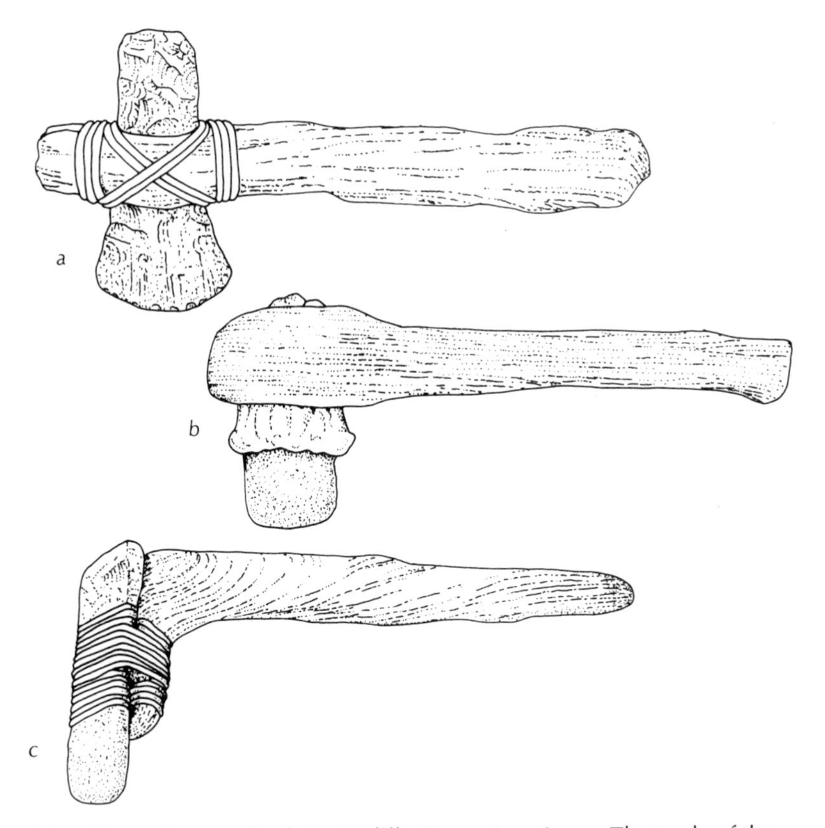

New Stone Age methods in Middle Stone Age times. The tools of the proto-carpenter were first shaped only by chipping, like this hafted ax (*a*), its retouched flint axhead lashed to a split-wood handle. The Mesolithic woodworkers learned that a chipped axhead cut better when its edge was ground down to a finer finish (*b*) and that handles lasted longer if an inset of antler, as seen here, served as a shock absorber. A sapling stem and branch (*c*) made a good adz handle.

dles, skis and runners for sledges. It is no mean trick to turn tree trunks into planking without the aid of a saw. Splitting is the only means available, and only trial and error will reveal which of the myriad forest trees splits cleanly and which does not.

Once this fact has been established — perhaps through

encounters with wind-felled tree trunks — the Mesolithic proto-carpenter had to plan ahead. Windfalls are scarcely a reliable source of raw material. The right trees would have to be chosen long before they were needed, chopped down with the new ground-stone axes, and left to "season" — that is, dry out — for a year or more before plank splitting could begin. We know something about the work investment of these proto-carpenters, again because of the thoroughness and imagination of Scandinavian investigators. Early in the 1950s a Danish team undertook to clear a two-acre space in a mixed-oak stand of the Draved Forest, using only flint axheads from the collection in the National Museum in Copenhagen. These they hafted in replicas of a famous wooden ax handle found in a Danish bog.

Two of the team, together with two professional loggers, undertook the axwork. The professionals could not reduce the force of their normal swings and thereby damaged several of the stone axheads. The amateurs, archaeologists both, found that the way to keep the flint unharmed was to use a short stroke, powered mainly with elbow and wrist motion. After a few days' practice they found they could chop down a tree more than one foot in diameter in about 30 minutes. Assuming that the amateur axman was steadily investing his muscular capacity of 55 foot-pounds per second during all the 1,800 seconds of the 30 minutes, this makes the work invested in felling such a tree the approximate equivalent of 17 horsepower.

The Danes also found that it was impractical to tackle trees much more than a foot in diameter. These they simply girdled, cutting away a ring of bark, leaving it to nature to fell the soon-dead tree. Such an approach must have suggested itself to the Mesolithic proto-carpenters; indeed, a need for forward planning of this kind suggests that the lumberers, hunters, fowlers, fishermen, and other specialists of the Mesolithic led sedentary lives in permanent encampments. Perhaps these fixed abodes were visited seasonally, a winter camp in one area and a summer camp elsewhere. Perhaps a single camp was used the year round for several years before exhaustion of local resources forced a move to a new site. What kind of evidence has been unearthed concerning the Mesolithic way of life in general?

Classical archaeology can be loosely defined as the study of the material remains of certain dead civilizations, mainly Greek and Roman, in Asia Minor, Greece, Italy, and those parts of the world, chiefly Europe and the Mediterranean Basin, that the Greeks and Romans colonized. Not so long ago students of classical archaeology were mainly concerned with only two subjects: inscriptions and fine arts. (This at least put them ahead of Biblical archaeologists, who were almost exclusively concerned with inscriptions.) Today their view is no longer so narrow. One consequence of the broader interests of modern classical (and Biblical) scholars is that we are learning more and more about life in the eastern Mediterranean before the age of metals. For example, the last decade has witnessed the discovery in Greece alone of three Paleolithic sites, four Neolithic sites, and one — on the shores of the Gulf of Argolis in the Peloponnesus — where, in a virtual layer cake, the remains of Paleolithic, Mesolithic, and Neolithic times lie superimposed. This bonanza is a site called Franchthi — if you say "Frank tea" you'll come close to pronouncing it — where various American and European field teams began work under the auspices of the Greek Archaeological Service more than a decade ago.

As a result of their labors we can now add to our knowledge of the fogbound Baltic and North Sea Mesolithic a picture of Mesolithic life along a sunny and, indeed, even desert-dry Mediterranean shore. Like their cultural cousins in northern Europe, the folk who lived at Franchthi between 8300 and 6000 B.C. were hunters and gatherers. For meat they slew the red deer, wild pigs, and sizable bovine animals that may have been bison or true wild cattle: the species *Bos primigenius*. They ate other animal food too: land snails, shellfish, and small fishes. They also collected plant foodstuffs: the produce of two wild-growing nut trees — almond and pistachio — and wild vetches, lentils, and peas.

Up until about 7250 B.C. they made their tools exclusively from flint and animal bone and sharpened the points of their bone awls by rubbing them on small grooved whetstones. For ornamentation they selected smooth flat pebbles that, after drilling a hole near one end, they evidently wore as pendants. Like their Paleolithic predecessors and their Neolithic successors they also made beads from that same tubular (and thus pre-"drilled") sea shell, the genus *Dentalium,* that was used decoratively at Mount Carmel.

Unlike their Paleolithic predecessors, they do not appear to have engaged in much funeral ceremony, at least if we may judge by the single Mesolithic burial uncovered at Franchthi. This was the body of a young male, aged about 25. The corpse, its knees drawn up toward the chest, had been placed on its side in a shallow depression and then covered with small cobbles. Neither ornaments nor grave offerings were present.

In the higher Mesolithic strata, dating from about 7250 B.C. to 6000 B.C., two notable and undoubtedly related new elements appear. First, in addition to the small fishbones that are evidence of some kind of coastwise fishing, the vertebrae of really large fish appear. The new bones are comparable in size to tunny vertebrae, and a typical tunny weighs upward of 200 pounds!

Second, a new tool material — the shiny black volcanic glass known as obsidian — enters the Mesolithic tool kit. The surprise in this may not be immediately apparent. Indeed, if certain British prehistorians had not recently traced down the principal sources of obsidian in the Mediterranean basin, no surprise would be possible. The British, however, did just that, and when the diggers at Franchthi sent samples of their Mesolithic obsidian off to them for analysis the source of the glass proved to be a volcanic deposit on the island of Melos, nearly 100 miles away.

The 10-foot-long fir dugout from Perth may have been suited to offshore fishing, but only a foolhardy crew would have paddled her 100 miles out to sea. At Franchthi no evidence at all has yet been found to suggest what kind of craft traveled the distance to Melos. Something other than a dugout? Even after the turn of this century, fishermen on the Adriatic island of Corfu went to sea in reed boats — although they did not go very far from their island base. But, too, only a foolhardy crew would try to pull one or more 200-pound fish into a reed boat.

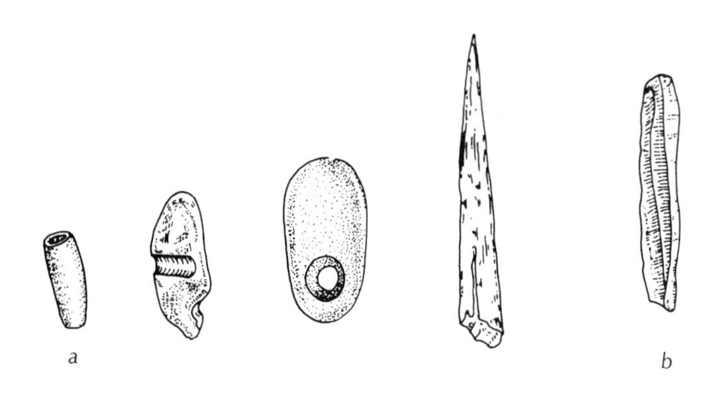

These artifacts come from the Mesolithic levels at a site on the Gulf of Argolis in Greece: Franchthi. Like the Natufians, the people here made beads of hollow *Dentalium* shell (*a*). Unlike the Natufians, sometime near 7250 B.C. they began to make tools out of a novel kind of stone: the volcanic glass known as obsidian (*b*). This gives the people of Franchthi the title of the earliest known seafarers. Why? Because the obsidian came from the island of Melos, thirty miles from the nearest point of land. They surely didn't swim there.

Thus it seems an inescapable conclusion that some efficient form of water transportation, quite possibly propelled by a sail as well as paddles, had become available in this corner of the Mediterranean late in the eighth millennium B.C. Maybe Melos was inhabited by an as yet undetected population of tunny fishermen who carried native obsidian with them to sell or trade along with their catch. Or maybe not. Maybe a Franchthi tunny fisherman, after a series of good catches had glutted the local market, decided to go exploring in his time off, and happened to stumble on the Melos obsidian. Or maybe not. All this is merely unsupported, not to say idle, speculation. The facts that remain are three. First, the obsidian found at Franchthi is dated at least 1,000 years earlier than all the other discoveries of this prized raw material in the Aegean basin. Second, the obsidian could only have come to Franchthi by water. The most direct route crosses nearly 100 miles of open sea and, if one instead skirts the shore and hops from island to island, the minimum open-water

TABLE FOUR

PERIOD	NON-MUSCLE ENERGY AVAILABLE	MACHINES
Interstadial through Final Würm Upper Paleolithic (Up until 12,000 B.P.)	FIRE	*Lever class:* digging stick, spear thrower *Wedge class:* bone chisel
Post-Würm Mesolithic (Up until 8,000 B.P., viz 6000 B.C.)	FIRE WIND? ALIEN MUSCLE: dog power	*Lever class:* As above, plus: paddle *Wedge class:* As above, plus: hafted ax, hafted adz *Rotary class:* bow drill *Storage class:* bow and arrow

passage is still more than 30 miles. Finally, the first appearance of obsidian at Franchthi coincides with an entirely different and very much increased capacity to catch fish. How would you connect all this up?

Before we leap flat-footed into the Neolithic, perhaps we should once again assess man's progress up to the sixth millennium B.C. I shall, as I did before, abbreviate the accomplishments of earlier times:

KNOWN MATERIAL CULTURE	NOT KNOWN BUT PROBABLE	EFFICIENCY 0.1 HP. x 12 hr. equals 50%
Small tools made on fine blades. Bone needles, toggles; tailored clothing. Projectile points of bone & antler. Decorated artifacts. Portable & semi-permanent shelters. Lamps, fired clay. Ceremonial pressure flaking. Ceremonial sculpture & painting.	Threads & cordage; snares & nets. Elaborate containers such as baskets. Practice of barter or trade.	on evidence of non-functional esthetic activities as an indication of increased leisure, increase to 75%
As above plus: Dugout canoes, sledges, skis, nets. Intensified plant food collection.	As above, plus: Plant husbandry? Sails?	on evidence of domesticated dogs and use of paddles for water transport, increase to 85%

This is no small series of advances. The bow as a device for storing muscle power; the bow drill, a machine utilizing the principle of rotary motion; the first proto-carpentry, with sharp-edged tools belonging to the wedge class of simple machines; the first attested water transport, aided by paddles, tools of the lever class; the first known improvements for winter travel, and, finally, the first domestic animal. If the Natufians were indeed selectively tending wild grasses, we could add incipient agriculture to the list. At Franchthi, when the Mesolithic population collects wild peas (in addition to wild vetches and wild lentils such as their Paleolithic predecessors gathered) and begins to harvest two entirely novel crops, wild almond and wild pistachio, we are also seeing something that verges on farm life.

Indeed, before this appears in print we may all have had to reset our sights as regards this point, aiming much farther into the past. Well up the Nile valley, not far below Aswan, investigators at a Palaeolithic site some 7,000 to 8,000 years older than Mount Carmel and Franchthi have recently discovered not only grindstones suitable for processing plant foods but also a few grains of the grasses that may have been processed. These prove to be of two kinds: barley and the primitive wheat botanists call einkorn. More of this plant material remains to be analyzed, and so it is now an open question whether the wheat was domesticated or wild (the barley appears to have been wild). The surprise, apart from the proof of a far greater antiquity for man's use of cereal foods than hitherto suspected, is that neither barley nor wheat is native to this part of Africa. Whether wild or entering upon domestication, the first seeds of both these grasses had to be transported up the Nile from somewhere else. How did this happen? Were humans involved? Did the seeds get a free lift in the bellies or stuck to the feet of migrating birds? Perhaps time will tell. For now we can only record it as another example of a milestone in human history lying much farther back on the road than anyone had anticipated.

What happens next?

CHAPTER
7

The Energetics of Crops, Flocks, and Pots

AT SOME time about 10,000 years ago — that is to say, in 8000 B.C.,give or take a millennium — a trend toward farming (already apparent in Egypt in Paleolithic times and in the eastern Mediterranean in Mesolithic times) became transformed into active agriculture and animal husbandry. The transformation took place independently in at least two widely separated parts of the world and involved unrelated plants and animals. The same transformation may also have begun independently in two or more additional parts of the world.

The two independent centers we are sure about are Central America in the New World and the Near and Middle East in the Old World. The earliest domesticated plants in the New World were squashes and gourds; they may have been cultivated in Mexico as long as 7000 B.C. and were certainly cultivated there by 6000 B.C. Thereafter the independent agricultural revolution in the New World brought under cultivation (to mention only the most notable plants) the unique New World cereal, maize, five different kinds of bean (the best-known being the Lima of South America), the white potato, the tomato, three world-famous tropical products — the pineapple, the avocado, and the papaya — and the source of an internationally accepted beverage and sweet, the cacao tree whose seed pods yield chocolate. For only a few of these plants is much known about the date of first domestication. For example, maize may have been cultivated after a fashion by 6000 B.C. and was certainly being cultivated by 5000 B.C.; the common bean and Lima bean were under cultivation by 6000 B.C.

The New World's contribution to global animal husbandry is, by comparison, almost trivial. The turkey is the best known animal; most of us are surprised to learn that the Muscovy duck is another domestic fowl of New World origin. After these two birds we can only add two closely related New World camels, the llama and alpaca, and a rodent relative of the porcupine, the domestic cavy or guinea pig. These latter witless pets seem to have been raised in South America since 2000 B.C. as a meat animal; as far as I know, few people any longer eat guinea pigs.

In the Old World, meanwhile, other innovative herders and farmers were busy. Not more than five miles from Shanidar cave (where the flower-strewn Neanderthal grave was unearthed) animal bones at an early open-air site were identified by the excavators to be from sheep and goats. In the lower levels of the site about one quarter of the bones of each species came from animals less than a year old. In the upper levels, which are dated at about 9000 B.C., the number of goat bones diminished

Worldwide agricultural origins are emphasized on this map, an extended polar projection. All the named crops except three have an annual yield in excess of 50 million metric tons; the crop name appears in the part of the world where the plant was first domesticated. Wheat, the largest of all crops (360 million metric tons) has been omitted; different wheats were probably domesticated independently in different areas. Two large crops, cassava (100 million metric tons) and grapes (60 million metric tons) were domesticated in two areas: cassava in lowland South America and Central America, grapes in the Near East and Europe. All 12 of the areas made some contribution to the Neolithic Revolution, although sunflower (North America), pigeon pea (India), and millet (Central Asia) are crops that are not produced in 50-million-ton quantities.

and the proportion of sheep bones from animals less than a year old rose from 25 to 50 percent. Evidently the people at Shanidar had taken control of the sheep population by then and merely hunted goats.

Fifteen hundred years later, at another Middle Eastern site

— Ali Kosh, in Iran — both sheep and goats appear to have been domesticated. Five hundred years after that, in about 7000 B.C. the people of Ali Kosh were raising two varieties of wheat and so were the people of Jericho, a mere trumpet blast or two away from the Mesolithic grass harvesters of Mount Carmel. The Jerichoans were also raising barley, lentils, and peas. Farther north, in Asia Minor, another group of eighth-millennium farmers at Cayönü was cultivating the same two varieties of wheat, along with lentils and peas, and was husbanding sheep and an entirely new domestic animal, the pig.

Evidently at about this same time the domestic potential of wild cattle was also recognized. The oldest evidence of their domestication known so far is from a site in Greece, Argissa, where the bones of cattle date to about 6500 B.C. Fifteen hundred years later, at Franchthi (the site in Greece that witnesses to Mesolithic seafaring) the full Old World quartet of domesticated animals — cattle, pigs, sheep, and goats — was being husbanded. Two thousand more years pass and the Neolithic period closes before the first evidence, in about 3000 B.C., of still another Old World bovid being domesticated. This is the familiar humped zebu, and the evidence is from a Mesopotamian site:

Animal domestication, a process with pre-Neolithic beginnings, took place independently in different Old World and New World areas. How long before European contact the Indians of North and Central America domesticated the turkey and those of Central America domesticated the Muscovy duck is not known. The Indians of South America, however, did their domestication long before the time of Christ. The sheep and pig had been domesticated in the Near East by 7000 B.C., and cattle by 6500 B.C. The horse was domesticated in Central Asia by 3000 B.C. and the water buffalo in India by 2500 B.C. The date of the reindeer's domestication in Northern Europe is not known. The two circled dots locate the two earliest finds of domestic dogs: Jaguar Cave in the New World (11,000 B.C.) and Palegawra Cave in the Old World (12,000 B.C.). The data on this and the preceding map are from Jack R. Harlan of the University of Illinois. The map projection is the work of Athelstan Spilhaus.

Tell Aqrab. At about this same time but many miles away in Central Asia some brave husbandman domesticated the horse.

All this gives the Neolithic population of the Old World a respectable barnyard full of mammals, but there were more to come. Out of Central Asia (perhaps around 1500 B.C., which is

very much post-Neolithic both in the Mediterranean and in the Far East) came the two-humped, or Bactrian, camel. From the same general area but at an unknown date the yak appeared. In a similar post-Neolithic setting (3000 B.C.?) someone in Egypt tamed the donkey. Sometime later an unknown donkey owner and an unknown but doubtlessly suspicious horse owner must have met somewhere between Egypt and Central Asia and made arrangements for the first mule. Just when and where this happened no one knows but mules were common work animals in Iron-Age Europe and, if the *Iliad,* the great Greek poem about the Trojan War, can indeed be traced back to the end of the Bronze Age, so can the mule, for Homer names and describes them.

We are nearing the end of the mammal roster. Around 2000 B.C. someone in Arabia or the vicinity domesticated the one-humped camel, the dromedary. (The name is not Arabic but Greek and testifies to the animal's fleetness of foot.) Except for reindeer, the water buffalo and its various odd Southeast Asian cousins, this completes the Old World list.

Still other domestics came to inhabit the barnyard, but our knowledge of their history is far poorer. Ducks of various species seem to have been independently domesticated in Europe, in Africa, and in the Far East. For example, the American market duck, the famous Long Island duckling, was first domesticated in China. Geese may have been domesticated independently in Europe and Africa or may have been introduced from the one to the other region in either direction. The commonest of barnyard fowl — hen and rooster — are somewhat easier to trace. Their obvious ancestral stock is the wild jungle fowl of Southeast Asia and so it seems most probable that they were first domesticated in that region.

It is that same region, by the way, that may have been an independent Old World center of domestication. It is certainly where Oriental rice was first grown and was also one of the two Far Eastern centers of pig husbandry (the other center is China to the north). Tea, the banana, citrus fruit, the mango, and that staple of the Pacific islands, the taro root, are other original

Southeast Asian domesticates. Chickens, pigs, and rice alone, forgetting all the rest, would have provided a solid foundation for a local Neolithic society. The fact that Southeast Asia supported a surprisingly early post-Neolithic society, characterized by a mastery of a complex metal, bronze, has been brought to light only in recent years. Its existence, hitherto entirely unsuspected by prehistorians, suggests a longer and richer past for the region. Just such a past might have been provided by a local, independent Neolithic revolution.

If we are reading the evidence correctly, farmers and husbandmen were firmly settled in many parts of the Old and New Worlds by 5000 B.C. Particularly in the Old World — between the eastern Mediterranean on the one hand and the western coast of India on the other — the village way of life had become commonplace and the first economic and political experiments that eventually gave rise to cities were already emerging. What, then, had the people of the Neolithic achieved as far as mastering energy (that is, advancing beyond their own muscle power) is concerned?

Foremost, they had captured the muscle power of truly large animals: cattle. Mesolithic or even Paleolithic man may or may not have harnessed dogs to sled and travois but that practice was and would remain a trivial use of animal power compared, say, to putting a yoke of oxen or even a single ox into harness. At first glance the domestication of animals may appear to be, like the domestication of plants, primarily a means of improving man's diet. A duck in the pot is worth two in the reeds, and so on. No doubt this factor was important; so was the ability to grow, rather than merely collect, plant food. We will see this farmer's increase in efficiency over simple hunters and collectors reflected when we assess man's Neolithic efficiency rating. Nevertheless, the use of animal power — the ox at the plow or at the sledge or perhaps even at one end of a lever, turning a millstone — was destined to lead to greater than simply dietary efficiencies.

It is not hard to see why. A farmer who takes a hoe and

prepares furrows for his seed has only his potential energy (one tenth of a horsepower) to apply to the task. What if, instead, he hitches an ox to a crude scratching stick? (Let me here decree that any ox worthy of its fodder has twice the work capacity of a horse and can therefore deliver power at a rate of 1,100 foot-pounds per second.) The answer to my question is additive: the man-plus-ox combination can deliver work at slightly more than 20 times the rate of man alone. Let us make an allowance for friction and say exactly 20 times. Given a yoke of oxen, the delivery rate rises to 40 times.

After the capture of large-animal muscle, what? Farming to one side, the most widespread Neolithic activity (and the one that is most abundantly evident to us today) was making use of clay. The people of the Neolithic pioneered a constructive variation on a human pastime that is immeasurably ancient: making mud pies. Their clayey activities I shall subdivide here into three unequal parts. The first can be called large-scale mud-pie play: the manufacture of adobe bricks for construction. (Adobe, incidentally, entered the English language as a loan word from Spanish, but the Spaniards had learned it from their one-time Arab overlords, who called a brick *Al Tob,* having discovered in Egypt that "tob" means brick.) For this first activity one may use any clay soil, the clayier the better, that becomes semi-fluid when dampened.

The second subdivision is potting: the manufacture of containers. For this activity one must use a respectable grade of proper clay. Proper clay is a water-loving silicate of aluminum, most commonly formed through the decomposition of feldspar, a mineral present in abundance in such rocks as granite and basalt. The most refined clays, the kaolins, form the raw material for the delicate kinds of pottery we call porcelain, but perfectly serviceable containers may be made from lesser clays, as is indicated by such names as earthenware and terra cotta ("cooked earth").

The final subdivision I shall call puttering. It includes, on

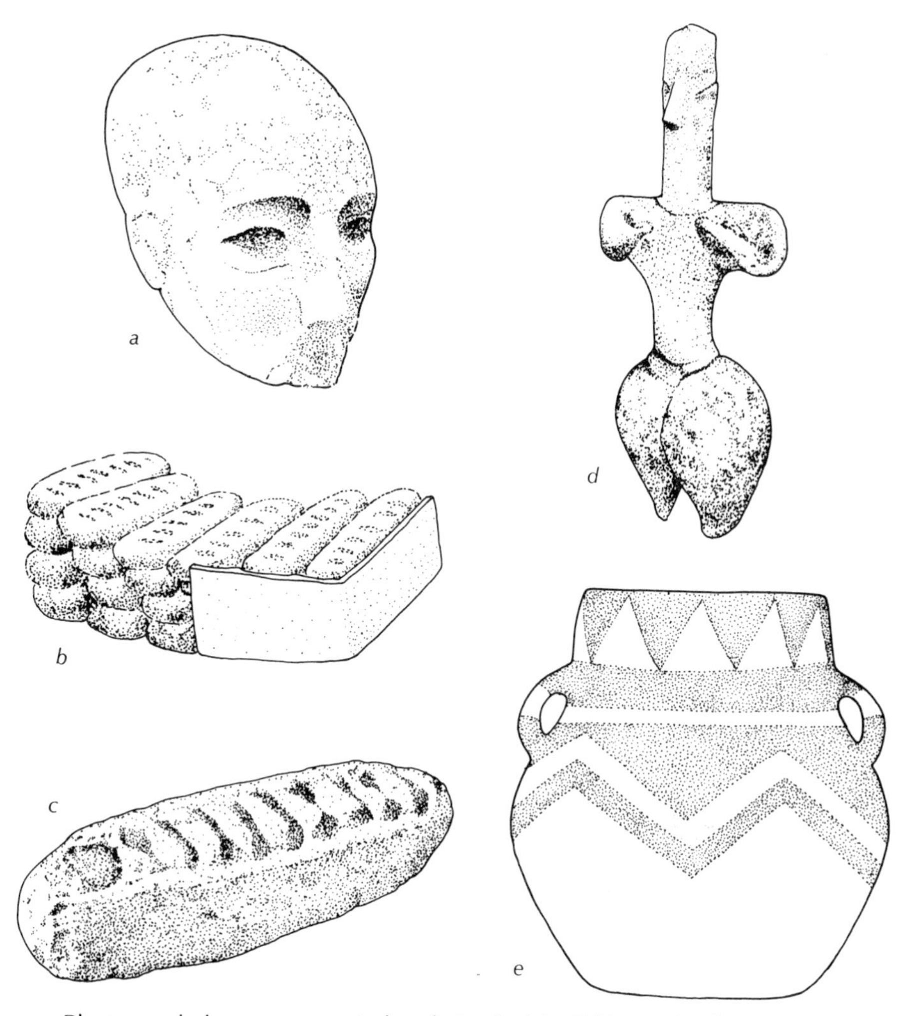

Plaster and clay were married early in the Neolithic. Indeed, recent evidence from Egypt indicates that plaster was being produced sometime before pottery first appeared. Some of the uses plaster was put to were bizarre. For example, at Jericho the undertakers sometimes covered defleshed skulls with plaster in imitation of the missing facial features (*a*). A more decorative use of plaster was to give a smooth surface to a wall (*b*) made of rather crude clay bricks such as the one seen here (*c*). Clay, of course, was used for much more than making bricks. Here, for example, is a small figurine (*d*) found at an early Neolithic village in Greece. First and foremost, however, clay was used to make pottery; the handsome pot seen here (*e*), with chevron decorations, was excavated at Jericho.

the one hand, the manufacture from clay of non-containers — figurines, writing tablets, sling missiles, and other often enigmatic objects — and, on the other, the prettying-up of mud-brick walls and floors with various kinds of plaster, some of them made of non-clays such as gypsum.

Of the three products, neither the bricks of the mud-pie player nor the non-containers and plasterings of the putterer have survived from Neolithic times in quantities that are in any way comparable with the broken (or occasionally intact) products of the potter. As any archaeologist will testify, the world contains a plethora of potsherds. Even though potters did not exist until some 9.000 years ago, the earth man has trod for thousands of thousands of years probably conceals more potsherds than it does stone tools.

How did it all start? We already know that the first clay figurines — in this classification, the work of putterers — were made thousands of years before the first potter went to work.

Who were the first potters? This is a "first" that is hard to pin down. What can be said with certainty is that, just as the first putterers were men of the Paleolithic, so the mud-pie players were Neolithic practitioners of plasticity well in advance of the potters. Houses with mud-brick walls were being built here and there all the way from Asia Minor (modern Turkey) to Palestine before the first pottery vessel appeared. Cheek by jowl with the mud-pie players, the putterers were also at work, making clay figurines and plastering mud-brick walls.

A particularly neat example of this temporal sequence is found at pre-Biblical Jericho. The earliest Neolithic levels there, which were occupied soon after 7000 B.C., contain mud-pie players' work: the remains of circular mud-brick dwellings. The occupants of those brick houses, called the "Stage A" people, are presumed to have been farmers (although non-farming villages, occupied by hunter-gatherers, are not uncommon in the Near East). The next Jericho occupants, the "Stage B" people, included mud-pie players and also had putterers among them. Their house plan was square rather than circular, and the floors were covered with a plaster made of lime, rather than clay or gypsum.

Some putterers, moreover, had gone into undertaking. The funerary practice of preserving skulls for secondary burial in groups began with the Stage A people. The Stage B people did this too, but added an extra touch; they used clay to "restore" the missing flesh of the face, sometimes even painting the clay and putting cowrie shells in the empty eye sockets.

Not until about 5000 B.C., or two long millennia after the first Jericho mud-pie players, did the first proper Jericho potter appear on the scene. Why so late? No simple answer comes to mind. If we are to judge from the appearance of pottery elsewhere in the eastern Mediterranean and the Near and Middle East, the Jericho potter was roughly 1,000 years behind schedule. Potters were at work in Asia Minor more than 8,000 years ago. They were present in Iraq and Iran soon after that and were at work at Levant sites not far from Jericho both a little before and a little after that same pivotal date: 6000 B.C. All we can say is that the Jericho potters were not the first to appear. Nor were they the last. Another 500 years would pass before pottery was made along the Nile, a river valley renowned as a major center of early civilization.

How do you make a pot? Until such time as the potter's wheel is invented — man's first significant advance in the application of rotary motion after the bow drill of the Mesolithic — you do just about what the first basketmakers did, using hand-rolled "ropes" of clay instead of bundles of grass. Instead of binding each successive rope to the one below with needle and thread, you let the natural plasticity of the wet clay do the binding. You help the process along by a little judicious mashing. This can be done by holding a smooth rock in your left hand, pressed against the inside surface of the pot, and whacking the matching outside surface with a wooden paddle. Done properly, this will not only squash all the ropes into a homogenized wall of clay but will also greatly reduce the thickness of the wall and thus the weight of the pot. If, this early in your career as a potter, you are also seeking aesthetic effects, you can carve the flat side of your paddle so that the outside of the pot will be decorated with dents. Even more easily, you can wrap cords around your paddle; they will leave decorative ridges in the clay.

When your pot has reached the desired size and shape, you simply let it stand until most of the water has evaporated from the clay. What you have then is an unfired pot. It is firm and fairly strong, but if you should leave it outdoors in a rainstorm it will once again turn into a mound of soft wet clay. Firing is the step that transforms the pot into a durable and no longer water-soluble container.

Mass-production pottery is fired in a special kind of oven that we call in English a kiln. (This was "cylene" in Old English, a direct borrowing from the Latin *culina,* which simply means "kitchen," and, by extension, "kitchen stove.") Whereas the first potter could have summoned his cousin, the mud-pier, and asked him to build an oven out of mud-brick, he and his descendants evidently did not think to do this for many generations. Instead they piled up their air-dried pots in a tidy heap in some open area. They probably put firewood among the pots and they certainly piled firewood up until it covered the whole heap. They may even have put green branches on top of the firewood to help retain the heat and piled loose earth on top of the branches, as the charcoal-makers of later times did.

In any event, the next step was to set the pile on fire. If the firewood had been stacked correctly it burned hot, soon reaching a temperature of more than 800 degrees Fahrenheit (450 degrees Celsius). The heat quickly evaporated the small percentage of water — not more than 3 percent — still present as residual moisture in the sun-dried clay. As the temperature continued to rise, additional molecules of water chemically combined with the clay were also driven out. This increased the density of the clay until it became no longer water-soluble.

Kiln temperatures in excess of 1800 degrees F. (1000 C.) are relatively easy to reach; clay raised to such temperatures ac-

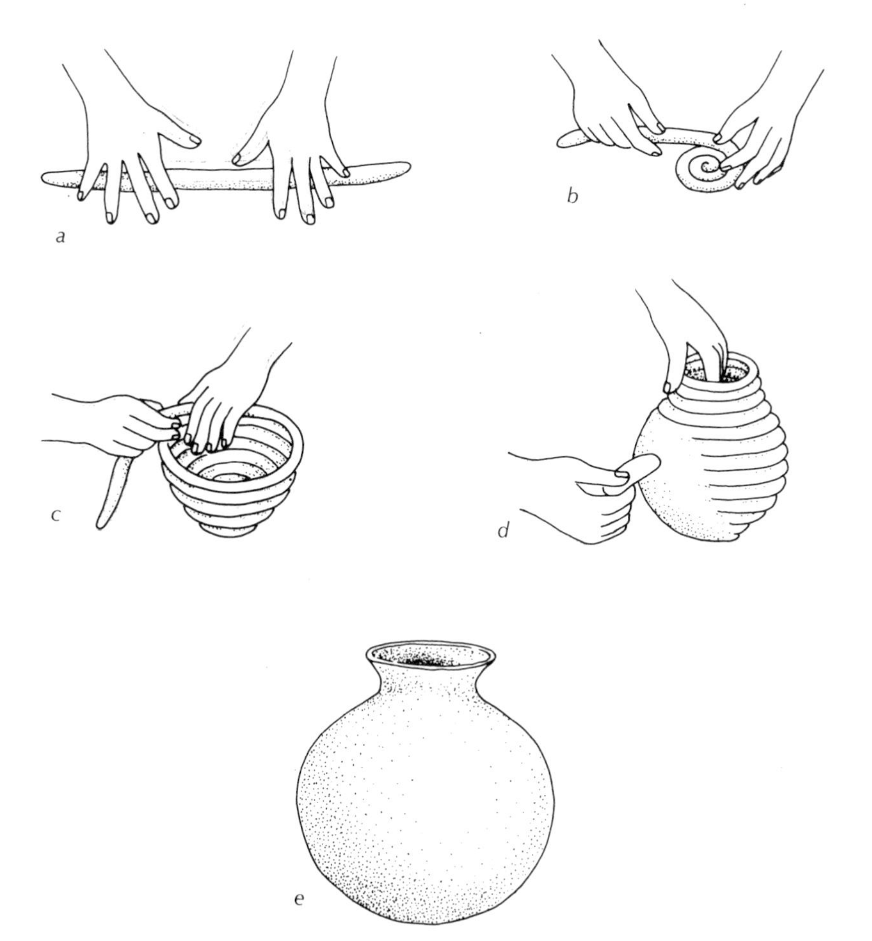

Look familiar? Very much like basketmaking, except that no thread and needle are required; the cohesive nature of the clay takes care of the binding. Start with a snake (*a*); curl it (*b*). When you run out of one snake, add another (*c*). After you have the shape you want, squash the snake edges together, outside and inside (*d*). (If you want to decorate it, slap the outside with a cord-wrapped paddle.) Now let your pot (*e*) dry in the sun until it is leather-hard. At this stage you can glaze it, carve it, add extras, or leave it plain for the kiln. When it is fired, chemical changes will make it waterproof.

tually fuses and becomes glassy, like porcelain. In the kind of open fire described here it is rare for the temperature to exceed 1400 degrees F. (800 C.), and even lower temperatures are not uncommon. One result of this is that most Neolithic pottery (and much of the common household pottery of later ages) is not very hard or particularly strong. This is one of the reasons why potsherds are so abundant!

Potting is not a unique invention. Clay is too common, too obviously plastic, and too easily fire-hardened by accident for one man in one place to have discovered the art and then diffused knowledge of it to all the rest of mankind. For example, the first pottery to appear in the Far East was made in Japan perhaps a full millennium before the first potter opened shop in Jericho; moreover, this was long before the inhabitants of Japan had even learned how to farm. So, just as there was a pre-pottery Neolithic period in the Near East, so there was a pre-Neolithic pottery period in the Far East. Perhaps this teaches us more about the shortcomings of labels like Neolithic than it does about man's progress.

In any event, the potters of Japan and also the potters of the New World seem to have invented the art for themselves. For the record, you should know that one small but respected school of prehistorians (they collect evidence favorable to diffusion, as opposed to independent invention) holds a contrary view. This school suggests that storm-driven fishermen from Japan, who somehow managed to survive a long involuntary voyage across the Pacific to South America, had among their crew some potters who introduced the art to the New World.

Nevertheless, now that we have potters among us — self-instructed or not — in a world that is settling steadily if slowly into the Neolithic rut of farming and stock-raising, just how has this changed the world? Principally by providing Neolithic producers with portable, vermin-proof, and comparatively leak-proof storage space. It is perhaps significant that all the crops brought under cultivation in the Old World before 6000 B.C. were grasses and pulses suitable for long-term dry storage. Wheat, barley, lentils, and peas could be stored in baskets or in pits dug in the floor of the house or even in pits and cribs outdoors, although such outdoor storage space is far from vermin-proof. Neither the olive tree nor the grape vine, however, comes into cultivation until sometime after potters appear on the scene.

I have already imagined that the Paleolithic collectors of plant foods probably observed the fact of fermentation and perhaps even enjoyed the results. Here in the Neolithic, with grain — the raw material for beer — available in abundance together with pottery vessels suitable for brewing beer and fermenting wine, it seems only reasonable to suppose that a new technician — the brewer/vintner — will soon be added to a growing population of Neolithic specialists. This same technician may also have dealt with olives: the technique of pressing olives to get oil is not much different from the earliest techniques of grape pressing, assuming the calluses on your feet are tough enough to let you tread on the olive pits.

The really crucial element in all this, however, may have been portability. No one can carry a storage pit from place to place. Baskets, too, are prone to leakage under the stresses and strains of travel. Skin bags, to be sure, are the equal of any pot when it comes to carrying dry produce in dry weather. If you want to keep dry produce dry in wet weather, however, you are best off with a pot. Carrying things from place to place, of course, suggests that trading, or at least exchange, was a substantial factor in Neolithic times. I believe this to be true. While some of the evidence for trade is direct, a good deal of it is only circumstantial, and the reasoning that leads to a favorable conclusion is roundabout at best. To follow me you will have to be patient but, as a reward, you will encounter some surprising facts that were unsuspected only a decade ago.

Let us first examine the direct evidence for trade. We have already seen that obsidian, the splendid volcanic glass that can be flaked into razor-sharp blades, was carried in Mesolithic times from the island of Melos in the Aegean to Franchthi, a hunting and fishing community on the Gulf of Argolis. The minimum overwater hop required to achieve this transport was more than thirty miles, and a point-to-point voyage would have crossed nearly a hundred miles of open sea. This is evidence of an unexpected maritime skill among the Mesolithic peoples of Greece and the Greek isles. Theirs is as nothing, however, compared with the maritime adventuring in the Mediterranean during Neolithic times.

Consider Khirokitia, a site on the island of Cyprus where the Neolithic villagers constructed elaborate two-story beehive houses using a combination of fieldstone and mud-brick. The

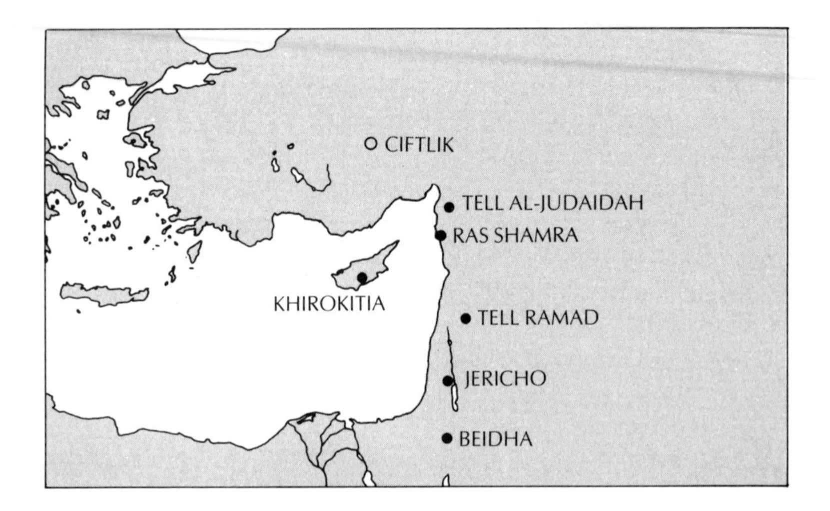

Trade in obsidian was common during the Neolithic. Analysis of trace elements shows that volcanic glass from one source in Anatolia (Çiftlik, open circle) reached five mainland sites to the south: (from the north) Tell Al-Judaidah, Ras Shamra, Tell Ramad, Jericho, and Beidha. To reach Beidha would have required an overland trip of more than 750 miles. Obsidian was also shipped overseas: the Çiftlik obsidian found at Khirokitia on Cyprus was carried more than 100 miles overland and then across 50 miles of open sea.

existence of this island village is fact number one. The shortest distance between Cyprus and the coast of Turkey is slightly less than 50 miles, and that is fact number two. In the middle of the fifth millennium B.C. a product of undoubted Turkish origin was transported overseas to Cyprus with some regularity. Want to guess what it was? Quite right: obsidian. As I suggested earlier, if obsidian is carried from an island to the mainland, as at Franchthi, explanations other than trade are possible. When, instead, as at Khirokitia, mainland obsidian is carried to an island, no explanation other than trade seems acceptable. What next?

Malta, a tiny island in the Mediterranean south of Sicily, contains a number of important archaeological remains, including some remarkable Bronze Age temples. Neolithic strata on the island include guess what? Right again: obsidian. It was once thought that the presence of obsidian at Malta was proof of contact between that island and Greek sources of the black glass. When, however, the Malta obsidian was analyzed (in the same way that the Franchthi obsidian was), it proved to have come instead from Pantelleria, a volcanic island 150 miles to the northwest, and from Lipari, another volcanic island, off the north coast of Sicily. Most of the voyage from Malta to Lipari can be made within sight of land; the 150-mile jump to Pantelleria, however, is no mean feat. Again the only acceptable explanation seems to be trade.

It is a fortunate accident that obsidian, a prized raw material for tools, is also not equally abundant everywhere. For example, one group of English investigators, including the now Disney Professor of Archaeology at Cambridge University, Colin Renfrew, has found that in the many thousands of square miles stretching from Asia Minor, the Levant, Mesopotamia, and the highland areas of Iran all the way south to the highlands of Ethiopia near the Horn of Africa, the sole sources of obsidian are in the northern and southern extremes. (The southern sources do not interest us here.)

Two of the northern obsidian sources are in Turkey and

the others are in Armenia. The black glass was carried from these sources to such Neolithic sites in Asia Minor as Haçilar and Çatal Hüyük (only 200 or so rough overland miles from the Turkish sources) and also reached Neolithic Jericho, more than 500 miles away. The Armenian obsidian traveled as far as the Turkish obsidian did, although in different directions. It is found near at hand at Çayönü, and also more than 400 miles to the south at Tell Ramad in the Levant and 500-odd miles to the southeast at Tepe Guran and Ali Kosh on the Mesopotamian border. How did all this obsidian move? How else but by trade?

Nor were the quantities trivial. The famous Neolithic site of Jarmo, in the Zagros Mountains of Iran, is some 300 miles away from the Armenian sources of obsidian. From the early strata at Jarmo, laid down well before 6000 B.C., come quantities of Armenian obsidian with a total weight estimated at 450 pounds. This is close to a quarter of a ton, which is no joke in an era before wheeled transport existed and even the availability of pack animals is not firmly established. To be sure, the 450 pounds did not arrive at Jarmo all in one piece, or even all at once as 450 one-pound pieces. Nonetheless, that is a lot of volcanic glass. The natives of the volcanic regions of Turkey and Armenia who made a business of supplying the raw material may even have found it a full-time business.

The implications of all this are weighty. As we have already seen, the Neolithic Revolution was a more or less simultaneous event worldwide. Particularly in the nuclear area of the Near and Middle East or, as some now prefer to call it, southwestern Asia, many archaeologists have labored long and hard in search of that elusive little settlement where the first wheat field was sown or the first flocks tended. Such searches will always be of interest, but their importance has very nearly vanished. With a network of interchange so widespread as the Neolithic obsidian trade, what Farmer Brown discovered yesterday would probably be told to Farmer Gray by some traveling salesman the day after tomorrow. Trade was trade, to be sure, but it was also communication and data accumulation, and, soon enough, dealing with numbers. Which, curiously enough, brings us back to clay once again and shifts us from direct to circumstantial evidence.

Or almost. Before we return to clay (and the origin of history) let us look at what the suppliers of the other great tool material — flint — have been up to. You will remember that the people who hunted in the vicinity of Poland's Holy Cross Mountains late in Paleolithic times had dug shallow shafts there in order to reach a particularly prized kind of stone: chocolate flint. This activity continued in Mesolithic and Neolithic times. Indeed, Neolithic flint mines have been found not only in Poland but in Western Europe and in the British Isles. What may have been a continuous but random activity of transient hunters during Paleolithic times in Poland had, however, been transformed into more than a casual activity in Neolithic times. A quick look at just one such flint-mining area in Western Europe will show us the difference.

Near Maastricht, in the southeasternmost tongue of the Netherlands, is a Neolithic flint-mining area so extensive that a decade of excavation has only managed to explore some 25 acres of it. The flint-bearing limestone here is covered by gravel and the kind of fine wind-drifted earth known as loess. At first the Neolithic flint-miners near Maastricht reached the limestone by the open-pit method. That is, they laboriously cleared away the overburden of gravel and loess and then dug into the soft limestone for the nodules of flint it contained.

By about 3000 B.C., a date established by carbon-14 analysis, the flint that was accessible by means of open-pit excavations had been exhausted. The miners then shifted to shaft-and-gallery excavation. The first step was to sink a narrow vertical shaft down through loess and gravel into the limestone. Hundreds of these shafts were sunk in the 25-acre area investigated thus far; the longest shaft was 60 feet deep and all were about three feet in diameter. Once the miner had reached the level of the flint-bearing limestone he tunneled into it horizontally. He then carried the flint nodules back to the shaft and a partner on the surface hauled up the flint.

The investigators of this flint-mining area estimate that the average yield of flint from each tunnel would have totaled about 10 tons. Given a grand total of 1,000 tunnels — a reasonable projection — this comes to some 10,000 tons of flint nodules. More to the point, if the total length of each shaft-and-tunnel combination is averaged at around 27 feet, the amount of spoil excavated from each would have been about 450 cubic feet, for a grand total of 450,000 cubic feet of loess, gravel, and waste limestone. Assuming an average specific gravity for this mix of 2.5 (limestone itself varies in specific gravity from 2.6 to 2.8), the total weight shifted was nearly 30 million pounds (15,000 tons). Every pound of this was moved by hand. If we consider only the average vertical lift (12.5 feet, or half the average shaft-and-tunnel length), the energy invested by the Neolithic miners to move the flint to the surface could have amounted to 250 million foot-pounds. Just shifting the waste inside each tunnel would have cost another 150 million. There is no use in attempting to convert the investment into horsepower because no possible way exists of calculating the time element in the equation. Thus, as to work done, one can only say that the flint miners of Maastricht did a heap of heaping.

What did they gain from their effort? The investigators estimate that 85 percent of the nodules the miners dug out of the limestone remained at the shaftheads in the form of spoil: trimmings removed in shaping various tools such as axheads and the

Flint-mining area at Rijckholt, in southern Holland, was exploited by Neolithic miners who drove hundreds of shafts, three feet in diameter, down through glacial gravels into a chalk formation that held nodules of flint (*a*). The miners then dug side galleries, a little less than three feet high, through the chalk where the flint was most plentiful (*b*), often breaking through into adjacent galleries (*darker areas*). In all, the hardy miners of 5,000 years ago may have removed 10,000 tons of flint nodules from this one area alone.

blades of adzes, as well as just plain waste. Another 4 or 5 percent of the remainder was used on the spot in the form of flint tools

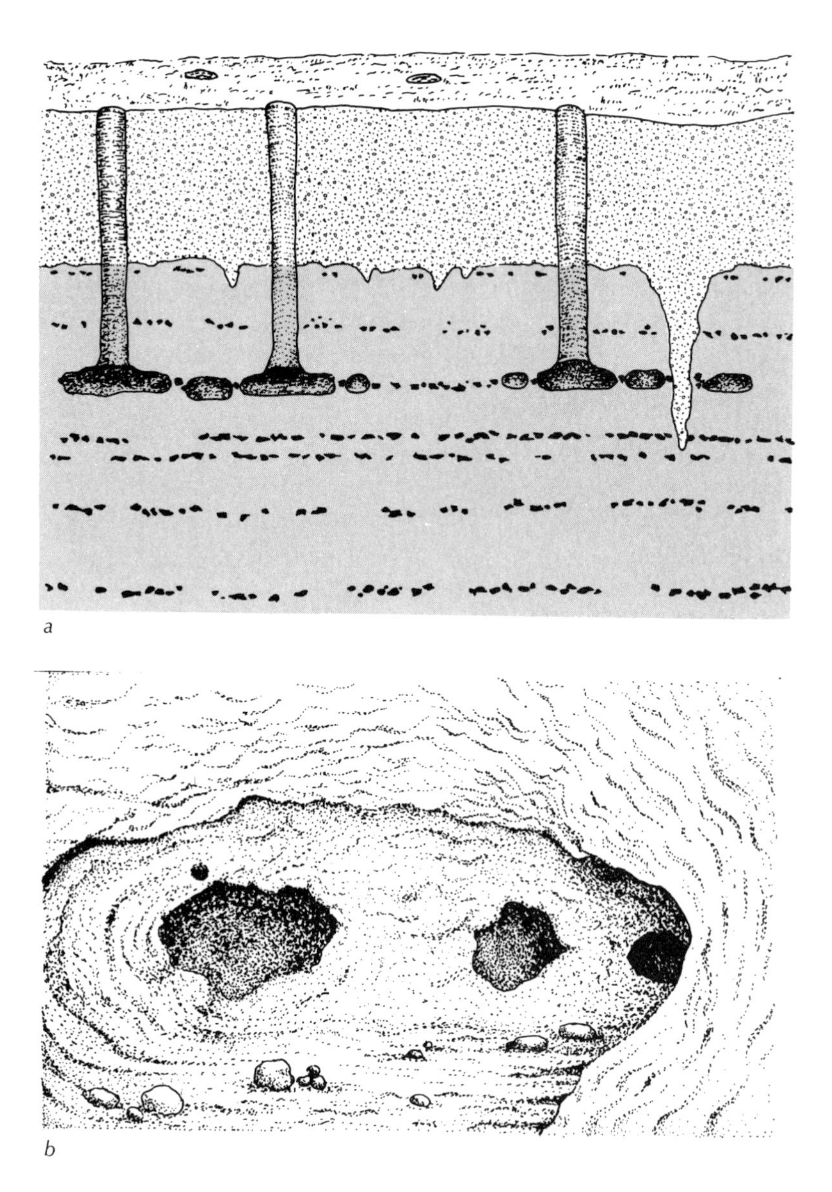

for the miners. This leaves a surplus of 10 percent roughed out to form axhead and adz-blade blanks; this flint was for trade. Tracing the movement of the Maastricht flints, the investigators have defined a trade net that extends at least as far as the middle reaches of the Rhine.

It seems reasonable to conclude that here, at least, flint mining was a skilled profession, even if not a full-time occupation. With the need for at least one worker down the shaft and another one at ground level, perhaps it was a family business, pursued by generation after generation of miners on a part-time basis. If so, the flint trade, too, might have involved family members. Question: Did the miners use any draft animals to pull the ropes that raised the flint, thereby reducing the amount of human muscle invested? We don't know. We can only hope to learn more about all this someday when other Neolithic flint mines are investigated. Meanwhile, back to clay.

Twelve years ago a young investigator whose interest was in the handiwork of the earliest mud-pie players and putterers began a review of various museum collections of man's earliest clay objects. These had been unearthed at various sites in Asia Minor, in the Levant, in other parts of southwestern Asia, and even at such widely separated places as Chanhu-Daro on the Indus River and Khartoum on the Nile. What she found, in addition to the mud bricks, beads, and figurines she had anticipated, was an incongruous miscellany of small clay objects that had puzzled or even bored their discoverers and that now came to puzzle her.

Neolithic sites of the ninth, eighth, and seventh millennia B.C. yielded these small objects, hand shaped out of clay of good quality and then baked to assure durability. They came in a variety of shapes: spheres that might have been marbles or sling missiles, ovoids, flat disks that could have been counters, cones and pyramidal forms that could have been game pieces, rectangles, triangles, crescents, and even odder forms. Many of the clay pieces were incised with bold lines, many had been punch-marked, and a good number had been pierced, as if for storing or for wearing suspended on a string loop, as amulets are.

The investigator, Denise Schmandt-Besserat, had studied at the Louvre Museum in Paris and was familiar with the collections from southwestern Asia that are housed there, collections that include material from sites where French archaeologists have been excavating for as long as a century or more. One such site, where work began in the 1880s and continues to this day, is the great Bronze Age city site of Susa, a proto-Elamite stronghold of the fourth millennium B.C. in Iran. At Susa, as at other city sites in the region, a system of accounting had been practiced that made use of small clay counters of different shapes — spheres, cones, disks and the like — that were sometimes packaged in hollow clay containers. In fact, it was a Louvre staff member, working with material from Susa, who first demonstrated the purpose served by these counters. It was not, however, until Schmandt-Besserat had compiled a catalogue of hundreds of her Neolithic clay "tokens," as she came to call them, that she realized how much they resembled the far later Bronze Age counters from Susa.

Her catalogue grew larger and larger. For example, from Jarmo, the 8,500-year-old obsidian-importing village, came a total of nearly 1,500 tokens: 1,153 were spheres, 206 were disks, and 106, cones. Schmandt-Besserat found that the earliest tokens as yet excavated were from sites in the Zagros region of Iran: the villages of Tepe Asiab and Ganj-i-Dareh Tepe, where the inhabitants may have tended flocks and quite possibly also grew grain in the mid-ninth millennium B.C. These simple incipient farmers made tokens in four basic shapes: spheres, disks, cones, and cylinders. Variations on these shapes and the use of added markings gave them a repertory of 20 different tokens in all. Tokens from a coastal site in Asia Minor, Beldibi, hundreds of miles west of the Zagros, are almost as old but include no cones.

Token-yielding sites occupied during the eighth and seventh millennia B.C. include three more points on the obsidian-trade network: Jericho in Palestine, Çayönü Tepesi near Lake Van, close to the primary sources of Armenian obsidian, and Tepe Guran in the Zagros south of Ganj-i-Dareh Tepe. The Jericho tokens include spheres and cones; the Tepe Guran tokens, disks, spheres, and cones, and the Çayönü Tepesi tokens, all four basic shapes. Not to burden you with further detail, the number of fifth- to ninth-millennium B.C. sites where tokens of one or more (usually all four) of these basic shapes have been found totals 27. Geographically, they are distributed from the shore of the Mediterranean (Beldibi) to the shore of the Caspian

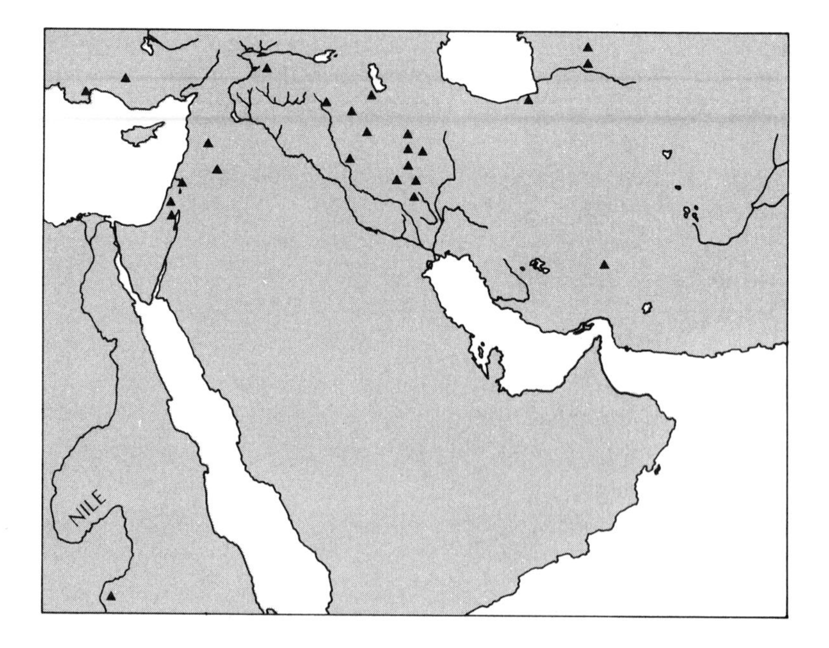

Neolithic sites, 24 of them in the Near and Middle East and one well up the Nile, where between 8,000 and 11,000 years ago clay "tokens" were made and probably used as counters in keeping track of animals and other items of barter. The Nile site is near Khartoum; the most easterly site is Tal-i-Iblis in Persia, and the most westerly Beldibi in Turkey. Many of the tokens resemble the earliest writing in the Near East, suggesting that one thing led to the other.

Sea (Belt Cave), and from the Iranian desert (Tal-i-Iblis) to the upper Nile (Khartoum).

Having come this far with me, you may ask impatiently, "What has all this to do with trade?" I told you that patience would be required, so follow me a bit farther. Remember the Louvre staff member (his name, by the way, is Pierre Amiet) who analyzed the clay tokens from Susa? He has turned up about 70 of them so far; their shapes include spheres, disks, cones, and cylinders. A typical Susa token container proved to hold one large cone, three small cones, and three disks; before being sealed inside the container each of the tokens had been pressed into the soft clay of the container's outer surface, as if to say: "Please find enclosed a big X, three small x's, and three y's."

Turn now to another Bronze Age site: Nuzi, in Mesopotamia, a city of the second millennium B.C. (by which time people had long known how to reckon and write). The palace archives at Nuzi were excavated in the 1930s and scholars are still studying the Nuzi texts. Also found during the excavation was an egg-shaped clay container marked with an inscription written in cuneiform. (This is to say, "wedge-shaped writing," because the writing implement made wedge-shaped marks in the soft clay. It would not be scholarly, however, to use whatever the word for wedge is in your native tongue and so the scholars use the Latin word, *cuneus,* instead). The Nuzi inscription listed 48 animals; when the excavators opened the container, out fell 48 tokens. Not long after this someone misplaced the tokens and so to this day we do not know what they looked like!

The Nuzi palace texts note the use of "counters" for bookkeeping; so-and-so many counters are spoken of as deposited, transferred, and removed. Let's think about this for a minute. The man who wedge-wrote the inscription on the egg-shaped clay container knew perfectly well that there was not even a pig's ear inside, let alone 48 pigs or goats or cattle or whatever. He was not tending flocks; he was keeping records. Because the container was found in the palace it seems likely that he was concerned with the royal accounts. But just down the street from the palace a private wedge-writer, perhaps an importer of lapis lazuli, might have been keeping his own accounts with the help of just such counters and clay envelopes. Imagine the scene.

Here comes a travel-stained young lad, all the way from Assur to judge by the cut of his cloth. He squats at the lapis importer's door, eases the pack off his back, and hands the proprietor a bundle and a clay envelope. The writing on the envelope informs the proprietor that the bundle has come from Assur and has been consigned to Nuzi, that the carrier is a nice young man (and family, too, a distant cousin), but that, while not meaning to reflect on anyone's character, the bundle should contain exactly as many lumps of lapis as there are counters inside the clay envelope. In effect, what the Nuzi importer has received is a pilfer-proof shipment accompanied by an unforgeable bill of lading. Far-fetched? Maybe.

A Neolithic trader of agricultural surplus, say pots of grain or heads of cattle, would also have found such a bill-of-lading system useful. In his case a leather bag of tokens would have served the same purpose the Nuzi clay envelope did; a record of transaction. If the Neolithic trader were a big merchant he might also find tokens useful in maintaining accounts: a string of disks could indicate how many pots of grain were stored upcountry or a string of hemispheres, how many cattle or sheep were in his herds.

Accountancy systems that make use of tokens simply as counters are well known historically. For example, our word "calculation" comes from the use of pebbles (Latin: *calculi*) in this way. Perhaps this is the most sophisticated use one can put counters to, short of the abacus. The very great diversity of the Neolithic tokens suggests, however, that the clay pieces were used not for arithmetic but to keep inventories and record transactions. Is there, in fact, proof that all of these tokens were not simply the game pieces or amulets or children's playthings many of their excavators took them for?

There is proof enough to satisfy me, and again Schmandt-Besserat is responsible for discovering it. Southwestern Asia is evidently the region where man first learned to write; the city-dwelling Sumerians of Mesopotamia seem to have been the first to practice the art and, as the fourth millennium B.C. drew to a close, among the Sumerian inhabitants of a city named Uruk were a number of busy scribes. Excavations in Mesopotamia have by now brought to light some 4,000 Uruk and Uruk-style texts using a repertory of some 1,500 signs, most of them abstract ideographs and some of them pictographs. (A pictograph is a more or less realistic sketch of the object it is intended to represent — for example, a drawing of a head with horns to signify an ox — whereas an ideograph is not in the least realistic.)

A few years ago, Schmandt-Besserat's collection of tokens included many made in Bronze Age times, a thousand or more years after the Uruk texts were written, and many others made in Neolithic times, three thousand or more years before the first Uruk scribes were born. She then began to compare the three-dimensional tokens with the two-dimensional Uruk signs. She found a number of parallels between the two groups. Her token catalogue is divided into 15 types, according to the shapes of the tokens; in 13 of these categories certain plain and incised tokens could be matched up with certain Uruk ideographs and pictographs. Eight of the matches are with known numerals, 10 are with Uruk ideographs of still unknown meaning, and a number match ideographs with known but rather abstract meanings, such as "peace" and "country." A number of others, however, represent concrete objects, and these, like the numerals, are eminently suited to accountancy, tax collection, and commerce.

There is, for example, a common token — a disk incised with a cross — that exactly duplicates the Uruk two-dimensional ideograph known to mean "sheep." Another disk, with a variant incision, closely resembles the Uruk ideograph for "ewe." The Uruk ideographs for "mat" and for "cloth" can also be closely matched by tokens, as can the rather pictographic Uruk sign for cattle. No fewer than four tokens representing different kinds of pottery containers (all intended for fluids?) are matched by pictographic Uruk signs. The meaning of one of the four signs is known: it signifies a vessel containing sheep's milk. The Uruk ideographs for wood, for metal, and for oil can also be matched with tokens.

Why should such a coincidence have come about? Schmandt-Besserat has put forward a possible explanation that incorporates the realities of Neolithic trade. First consider that a repertory of token shapes and markings was known throughout southwestern Asia over a span of some 7,000 years, almost from the dawn of the Neolithic until well into the Bronze Age. In several instances, tokens were tucked inside clay envelopes, the

Six tokens, the three to the left in each of the two pairings, are compared with six tablet markings of the kind used in the earliest known writing, the mixture of pictographs and ideographs used at Uruk in Mesopotamia early in the Bronze Age, before cuneiform writing was invented. The meaning of the six Sumerian words is known. They are, from top to bottom (*left*) dog, mat, and metal and (*right*) cow, sheep, and wood. The likeness in shape and markings between the three-dimensional tokens and Sumerian writing suggests that the widespread use of tokens from Neolithic times onward inspired some clever record-keeper to become the first scribe.

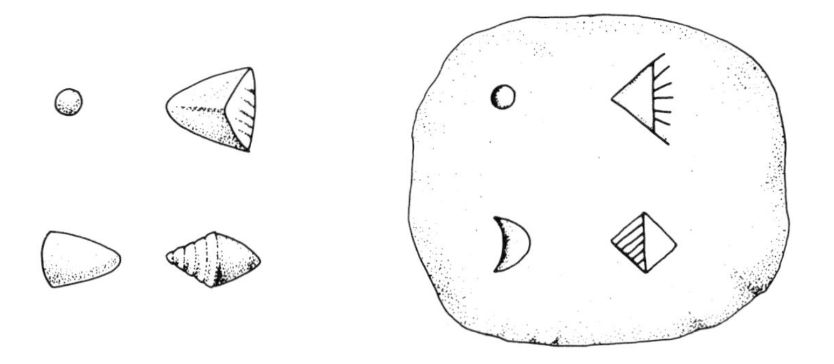

How it may have happened: instead of sending four tokens with a shipment of goods, to act as a bill of lading, a merchant could instead have punched and scratched their images on a flat clay tablet and sent the tablet instead. The four Sumerian words shown here are (*top*) "10" and "cloth" and (*bottom*) "60" and "bracelet."

outside of the envelope was marked to indicate its contents, and the envelope then changed hands. Let us hypothesize that such a practice was common and that it became even more widespread as Neolithic village communities gave way to the first urban settlements of the early Bronze Age.

The scene that follows is not to be blamed on Schmandt-Besserat. Its imagery is all my own. We see a busy Uruk merchant, searching for a simpler bill-of-lading system. "Why," he asks himself, "do I have to send and receive hundreds of tokens when a properly marked envelope would do just as well all by itself?"

"Let's give it a try," he exclaims. "What am I sending upcountry this week? Ten bolts of cloth and 60 bracelets. That's 10 pyramids and 60 double-cones if I use tokens. But what if . . . Here, master potter, lend me a handful of clay. I'll handle the numbers another way. Here's the token for the number 10; it's a small sphere. All right, I'll poke a circular hole in my clay to represent the number 10 and beside it I'll scratch the outline of a triangle to represent the pyramid that means cloth. Now, here's the token for the number 60; it's a cone. So I'll poke a conical hole in the clay to represent 60 and beside it I'll scratch the outline of the double-cone that means bracelet . . . Oh ye gods, I've just invented writing! I say, how very convenient for us merchants. Hmm; that fat cousin of mine, the tax-collecting priest who's always complaining about his token-calluses, I'll bet he'd be interested too. . . ."

Preposterous? Maybe. But these facts remain. Ideographs far outnumber pictographs in the Uruk texts. Such texts as can now be read are remorselessly concerned with transactions; among the arbitrary Bronze Age signs that appear most frequently are those for cloth, sheep, cattle, beer, and bread, and the forms of many such signs resemble the forms of Neolithic tokens. Last of all, as the art of writing began to spread in the fourth and third millennia B.C., the use of tokens seems to have slacked off. For example, the very many subtypes of tokens common in the fourth millennium had been reduced to only four or five by the end of the third millennium.

But we are venturing too far into what still lies ahead. The primary point I am trying to make here is simply that very much more trading in a number of commodities went on during Neolithic times than just the well-documented obsidian and flint trades, and that Schmandt-Besserat's catalogue of Neolithic tokens strongly hints at just how much more this was. If the token system also inspired the development of writing, without which history cannot exist, so much the merrier.

How shall we summarize the Neolithic? Settlements were few and scattered when the New Stone Age began; when it drew to an end much of the world was filled with hamlets, villages, and even small towns. The man-of-all-work now was not the hunter but the farmer. Hunting as a way of life continued only on the fringes beyond the plowed land; where the Neolithic Revolution had made its mark the hunter was only one among an array of new craftsmen; potter, weaver, tanner, brewer, baker. (All five crafts, by the way, remain echoed in common family names to this day.) In this connection, those proto-carpenters, the Woods and the Wrights, were probably established by now, but the Sawyers and the Carpenters were not destined to reach craftsman's status until, assisted by the Miners, the Age of Metals had been ushered in — by the Smiths, naturally.

From the viewpoint of energy there are other points to be made. Certainly the major muscular achievement of the Neolithic was the domestication of large animals suitable for draftwork; the plow ox to begin with, and, eventually, horses, donkeys, and the horse-donkey hybrid, the mule. The latter two work animals, as noted, were post-Neolithic domesticates. So, for that matter, was the horse, if one reckons by an absolute time scale. But the first Central Asiatic horse breaker in 3000 B.C. was still living in Neolithic times in his native heath, just as were most New World peoples until the 16th century A.D.

Leaving aside the net gains in efficiency that are represented in herding rather than in hunting, and in farming rather than in collecting, perhaps next in rank as a major Neolithic achievement is the new application of fire. Let's think about this for a minute. When man first learned to capture and tend fire his gains were fourfold. First, he had light at night. Second, he had a way to keep warm in cold weather. Third, he could begin to cook his food. Fourth, he could use fire to alter materials: fire-harden the tip of a wooden digging stick or a spearpoint and even straighten a bent branch to form a better spear shaft. This was about as far as the uses of fire went before Mesolithic times.

When we pass on to the Mesolithic we see that fire is being used in the same four ways. Somewhat more ambitiously, to be sure, as far as altering materials is concerned; consider the Scottish dugout, hollowed by repeated small fires. And possibly somewhat more frequently too; consider the bending of sled runners and skis. But still, fundamentally, these ways of using fire are still the four old ways.

The first Neolithic potter changed all that. The four old ways to use fire continued, to be sure, but what the potter did when he fired his first pot was to alter a material in a new and quite different way. The heating affected the chemical composition of the clay, changing it from a water-soluble material into a waterproof one. The fire-hardened portion of a spear point will eventually wear away with use, leaving its owner with an ordinary stick once again, but there is no way to turn a fired pot back into a lump of clay.

The potter's step is an important one in ways that have nothing to do with the usefulness of his pots. It marks a realization that other common earth substances, not simply clay alone, can be radically altered by exposure to fire. This realization opened the way to the Age of Metals.

As we have done before, from time to time, let us again review man's accomplishments, this time carrying the date forward to late in the fourth millennium B.C., when southwestern Asia, at least, was on the threshold of the Age of Metals. As you will see, this is our fifth (and last) Table.

TABLE FIVE

PERIOD	NON-MUSCLE ENERGY AVAILABLE	MACHINES
Post-Würm Mesolithic (Up until 8,000 B.P., viz 6000 B.C.)	FIRE WIND? ALIEN MUSCLE: dog power	*Lever class:* paddle *Wedge class:* hafted ax, hafted adz *Rotary class:* bow drill *Storage class:* bow and arrow
Neolithic (Up until 5,000 B.P., viz 3000 B.C.)	FIRE WIND: sails ALIEN MUSCLE: As above, plus: draft oxen horses	As above.

KNOWN MATERIAL CULTURE	NOT KNOWN BUT PROBABLE	EFFICIENCY 0.1 HP. x 12 hr. equals 50%
Dugout canoes, sledges, skis, nets. Intensified plant food collection.	Plant husbandry? Sails?	on evidence of domesticated dogs and use of paddles for water transport, increase to 85%
As above, plus: Plant cultivation, animal husbandry. Pottery, brick structures. Trade and reckoning.	Beer. Wine? Experimental metallurgy.	domestication of draft animals and use of sails puts this index over 100% and renders it meaningless here and hereafter

CHAPTER

8

Metals, Rotary Motion, and Wheels

MAN'S NEXT conquests tie together in a manner much like that of the old nursery rhyme about the stubborn pig. The herder, his patience lost when the pig refuses to jump over a stile, summons his dog to bite the pig and, when the dog refuses, commands a stick to beat the dog, and so on and so forth until the infuriated herder's final command went something like this. "Butcher, butcher, kill ox, ox won't drink water, water won't quench fire, fire won't burn stick, stick won't beat dog, dog won't bite pig, pig won't go over the stile, and we shan't get home before morning!" The butcher started, and so did everyone else.

If we can imagine some great ancient innovator talking to himself alone on a windy hilltop, brooding before the dawn of the Age of Metals, his words might have had a similar, although more negative, ring. Hear him cry out, "Miner, miner, find ore, no ore, no metal. No metal, no saws. No saws, no wheels. No wheels, no mills. No mills, no power trains. No power trains, no . . . Whew! Hold it there; I'm ahead of my times." And so we shall hold it too, heeding for now only the innovator's initial command, rather than the consequences of disobedience.

How does one recognize metal? Before you learn what metal is, you don't. The metals that come in nugget form simply seem to be other kinds of rock. Since you have now accumulated a couple of million years' experience with various kinds of rock, these oddities are not likely to command your attention. There are, after all, lots of odd rocks. Soapstones, for example, are so easy to cut that you can carve big chunks into splendid pots using nothing but a flint or obsidian tool. Many limestones, too, can be scratched with nothing harder than a fingernail. So what's so special about these soft yellow or reddish pebbles? They bend, that's what! And they're the only rocks that do!

So now you know what makes these rocks different from all other rocks. What are you going to call them? In later centuries, when the miners' search for ores was a full-time endeavor, the Greeks developed a suitable name. The Greek word for "search" is *metallao,* so it was easy enough for the Greeks to call the action of prospecting for ores *metalleia*. From that point it was only a short step to call a mine *metallon.* This the Romans transformed into *metallum* and the sensible French shortened to *metal,* a word they applied both to the mine and to what was mined. By the way, the English word "mine," scattered about so liberally here, is neither Greek nor Roman in origin; it seems to be Celtic, closely parallel in its sense to the Greek *metalleia,* and meaning "to dig ore." In any event the English word "metal" comes from the Greek via the French and is thus fairly late in origin. The anachronism need not trouble us. Whenever the first protosmith whacked the first soft nugget of gold or copper into some shape that suited his fancy, the Age of Metals began, even though hundreds of centuries had to pass before the modern name for these deformable stones was invented.

By sheer geological chance the four most common metals found for the taking in "native" (or nugget) form are all soft enough to be bent by hand. What are they? Three of them you can readily guess: gold, copper, and silver. The fourth, a far more humble metal, may come as a surprise to you. It is lead, the element abbreviated as Pb by technical folk. The abbreviation is of a Latin word, *plumbus,* which you will recognize as the root of the proper craft-name: plumber. (Until a century or so ago most plumbing, that is to say water pipe, was made of lead.) Of the four soft metals copper is the most abundant in native form, followed by silver as a very poor second and by gold as an even poorer third. Native lead is the scarcest of all, but the shiny lead ore, galena, is the most easily smelted of any metallic rock.

One can declare without fear of contradiction just what the worldwide distribution of the first three native metals once was. This is because, barring some gold nuggets in unworked gravel deposits here and there, none of them any longer exists in native form. All the easy-to-find surface and near-surface nuggets were long ago picked up and put to use. Estimates have been made, however, and these suggest that the three were not

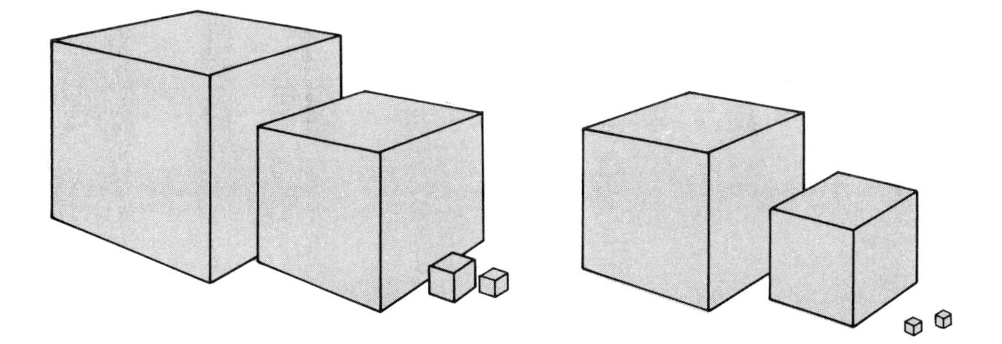

The abundance of four "native" metals, estimated for the Old World (the prehistoric Near East) at left and for the New World (the pre-Columbian Andes) at right. Three of the metals, copper, silver, and gold, were available in nugget or in other nearly pure forms. The fourth, pure lead, is rare, but the most common lead ore — galena — is easily smelted and was probably the first ore so treated. The largest cube at the left represents native copper: for every 1,000 pounds of native copper in the Near East there were (in descending order) enough lead ore to furnish 400 pounds of pure lead, five pounds of native gold, and two pounds of native silver. The largest cube at the right represents pure lead; lead ore was about as abundant in the Andes as in the Near East. For every 400 pounds of lead there were 200 pounds of copper and one pound each of gold and silver. Both native copper and native gold were far more abundant in the Old World than in the New World.

equally abundant everywhere. In southwestern Asia, whence my imaginary innovator cried out, for every 1,000 pounds of copper nuggets there were perhaps five pounds of gold and only two pounds of silver. This proportion of native gold to native silver is unusually gold-rich, a fact that may well have had some effect on the prehistory and ancient history of the region. For an opposite example, in much of the New World — where metallurgy, like agriculture, was independently invented — native copper is only one fifth as abundant as in Asia, while gold and silver are about equal in rarity and only half as abundant as silver was in southwestern Asia.

Who was the first to discover the malleable nature of metal nuggets? To answer that question with any certainty would seem to be unimaginable. We can, however, respond to the "where?" question that follows by saying, "Somewhere in southwestern Asia," and to the "when?" by saying, "Sometime during, or even quite early in, the Neolithic." There is even, for all its unimaginability, one present candidate for the title of "who." This is the person, age and sex unknown, living in eastern Asia Minor in the seventh millennium B.C., who hammered on a bit of native copper until the soft metal was shaped into a crude kind of pin, about half an inch long and a sixteenth of an inch in diameter. The hammering was no accident; the nameless proto-smith made not one but three such pins and sharpened their ends. The three pins, the earliest known metal artifacts, were unearthed at Cayonu, a site in a mountainous part of what is now the frontier between Turkey, Syria, and Iraq, by a joint Turkish-American team of archaeologists.

But was the proto-smith of Cayonu really the first to work with metal? Surely not; encounters with malleable rock such as his must have been commonplace round the world. There is even a New World example — of epic proportions although far later in time — of what simple stone-tool users did when given copper as freely as the Greeks thought Prometheus had given mankind fire. Once again sheer geological chance was at work, this time on the southern shore of Lake Superior, greatest of the Great Lakes. There, in deposits bared by the surface scouring of the last great ice sheet, was an abundance of copper free for the taking. Not copper in nuggets, but great flat sheets of native copper, ready to be peeled away and chipped and hammered into any shape the New World proto-smith desired. At some time 7,000 to 6,000 years ago, when the Indians of Central and South America were already busy with agriculture, a few game-hunting Archaic Indians of North America came upon this copper bonanza. What they did with it has caused them to be known today as the "Old Copper Indians."

The Old Copper Indians made chisel-like narrow axheads out of the native copper. They made lance points, some with narrow tangs for hafting and others with half-sockets. They made any number of knife blades, some in the semi-lunar Eskimo slate style and others with long tangs for hafting. Had they been farmers like their cousins far to the south they would also doubtless have made hoe blades for their cornfields. But they weren't farmers; in fact, they weren't even proper smiths. They

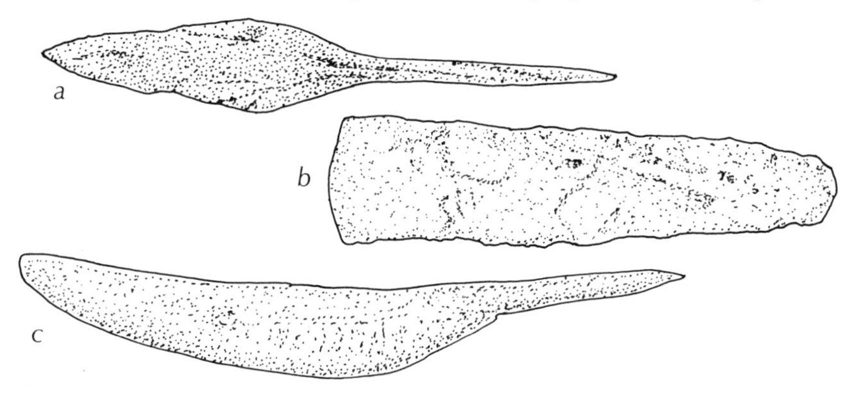

Unsmelted copper tools: these three artifacts — a spear point (*a*), an axhead (*b*), and a knife (*c*) — are typical of the metalwork done by Archaic Indians of the Lake Superior area some six to seven thousand years ago. They were metalworkers before metallurgy had been invented in the New World because they lived near great deposits of native copper in the form of sheets free for the taking.

did bury their dead, however, and it is in their burial grounds that many of these copper artifacts have been found.

In effect the Old Copper Indians had been inspired by the easy availability of this native sheet copper to invent a kind of specialized metallurgy. They never smelted or hot-welded, as their civilized compatriots in Central and South America learned to do, using nuggets of gold, silver, copper, and even platinum. But they did apply heat; when one of their copper pieces had been turned brittle by overpounding, they heated it over a fire until it became soft again. This is the metallurgical process known as annealing (the word is of Old English origin: *on,* meaning "put in contact with," plus *aelan,* "burn, or fire"). They could also cold-weld one edge of a sheet to another by hammering.

As far as the smiths of the Old World were concerned, these and the later, more advanced metallurgical feats of the civilized New World (including the art of casting and the independent invention of bronze) were already old hat by the fourth millennium B.C. Whether or not the pioneer Old World metallurgists had heard the bidding of my phantom innovator, they passed beyond the stage of dependence on nuggets for their raw materials more than 5,000 years ago. Their hand-in-glove associates, the miners, sought out ores and not nuggets, and they were just as prepared to invest foot-pounds by the millions to find ores as their flint-mining forebears had been.

The difference between a nugget and a lump of ore is easily explained. The nugget, being native metal, is virtually pure. Nuggets of copper (and the rare nugget of lead) usually contain almost no impurities and that slight impurity in copper is usually a trace amount of silver. How can that hurt? Native silver is at least 99 percent pure. With native gold the level of purity is lower: about 90 percent. The impurity in gold is often silver, which is why one of the precious metals of southwestern Asia and the Mediterranean was called *electrum,* from the Greek word for amber. This natural mixture of gold and silver has a tawny, off-yellow color not unlike the color of amber.

In contrast to this, ores are mixtures of mineral elements, many of them not even metallic. The proportion of the desired metallic element to the rest, the "waste" elements, can be very low. For example, it is not at all unusual, today, to mine and process ores that contain far less than 5 percent metal. The key word in that sentence is "process." Let me suggest a memory aid; melt versus smelt. In order to liquify native metals, we heat them until they are *melted*. In order to liquify the metals locked up in ores, we heat the ores until they are *smelted*. On that prefixed letter "s" is suspended quite a lot of technology.

Why liquify? So that the metal may be cast — that is, poured hot into a prepared mold and then allowed to cool until

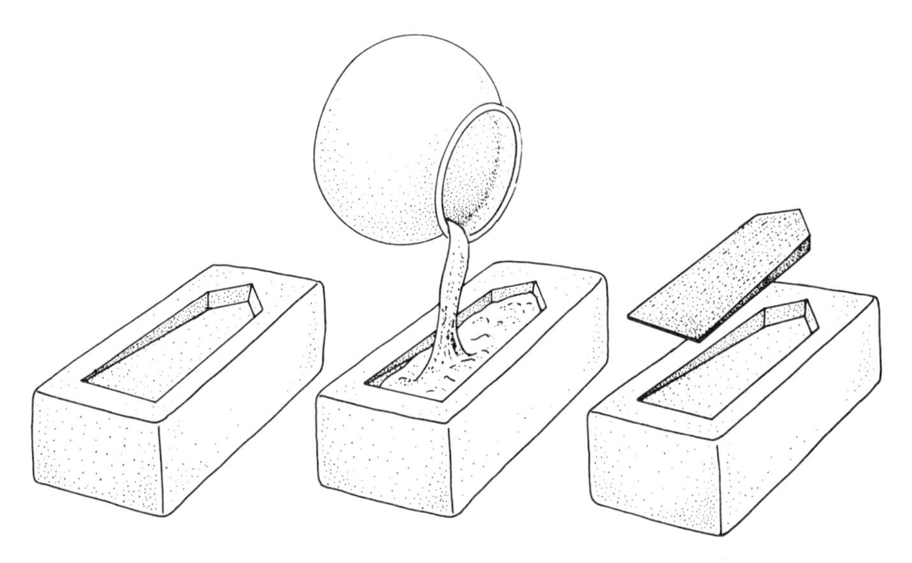

Early standardized production: with a fire hot enough to melt copper and a mold made of stone (with a cavity of the desired shape laboriously pecked in it), the smith need only pour the molten copper into the cavity, wait until the metal cools, pick out the cast object, and pour the next one. Here a plain ax blade is being cast; the soft copper will be work-hardened later by hammering. Bronze, an alloy of copper and other elements, is so much harder than copper that, except for decorative uses, pure copper artifacts went out of fashion.

solid again. (The mold, made perhaps of clay or stone, has been shaped in the negative form of the solid metal object you wish to produce.) When the liquid metal solidifies, you have before you the article of your desire: axhead, adzhead, hammerhead, sword blade, dagger, arrowhead, spearpoint, bracelet, necklet, or what you will. The hot liquid? Silver and gold for the ornaments. Bronze (the first of the great man-made metal mixtures — alloys) and copper for the tools and weapons.

Am I going too fast? Probably so. Let's backtrack for a moment. The simple art of melting, which allowed all these feats of casting, would not have been within Mr. Smith's grasp had not Mr. Potter and his Neolithic forebears developed the high-temperature technology and the kiln-proof pots Smith needed to liquefy his nuggets. Also, somewhere along that earlier line, the potters became interested in adding fine colored glazes to their wares. This may have further whetted the miners' interest in certain attractively colored rocks, already in some demand among the ladies for grinding into cosmetic powders or shaping into beads. Two of the richest copper ores, azurite and malachite, are respectively handsome blue and green stones.

I am not going to argue about which ore was the first to be smelted, although a good case can be made for galena (to produce lead). As to the discovery of the first great alloy, bronze, experts will be arguing the fine points of this advance for generations to come. Even now, when many lines of evidence remain unexplored, it seems probable that bronze (an alloy of copper with certain other elements) was independently discovered at least three times: Once, as we have seen, in the New World, rather late on; once in southwestern Asia, probably the earliest discovery of the three; and once again in southeastern Asia, at an early but uncertain date. At least in southwestern Asia the discovery may better be described as learning by experience.

This is because copper ores exist in a remarkable range of forms. Some of these contain arsenic. Smelting such an ore produces the natural alloy known as an arsenical bronze. There are even ores that contain both copper and tin; smelting them produces the best-known of the bronze alloys: tin bronze. Will you be surprised to learn that the addition of lead to copper also produces a kind of bronze?

Thus it is not so much a question of who discovered bronze but rather a question of how many dim-witted coppersmiths failed to realize they had done so accidentally. Again, it is doubtful that we shall ever learn the answer. I can, however, add a grim footnote. Those who were smelting the ores that produce arsenical bronzes probably all died early; the fumes from such a smelter are extremely toxic.

The oldest certain site of copper smelting is at Tepe Yahya, in Iran. It is nearly 6,000 years old, so we can see that metallurgical progress was swift following the day the pinmaker set up shop on the Turkish border. Copper mines that are still older are known in the Balkans, but what the miners did with the ore is uncertain. They could have been digging out malachite for nonmetallurgical purposes. In any event, smelting at first seems to have been a simple business. Its methods are unique but its components — a kiln-like smelting furnace and charcoal for fuel — had been known to potters for many centuries.

The coppersmith would have begun by building a small-scale furnace (with air vents near the bottom) that differed from a potter's kiln only in being open at the top. No great amount of effort was put into smelter construction because, in these early days, each smelter was likely to be used only once. A chimney-like cylinder of fieldstone, chinked with mud, would serve the purpose. To charge his smelter the coppersmith first put some dry wood at the bottom, where it could be set on fire through one of the bottom holes. On top of this kindling he added alternating layers of charcoal and copper ore; as he grew in wisdom he learned to mix some kind of mineral flux, such as limestone, (the Latin root, *fluxus,* means "flow") with the ore to make the metal run. When the chimney-like smelter was filled to the top the coppersmith touched off the kindling and either let the natural draft increase the temperature in the smelter or forced air into the smelter through the bottom holes to achieve the same effect. In a while the smelter became about as hot as a potter's porcelain kiln.

The rising temperature transformed the charcoal into superheated carbon gas. When the smelter temperature edged above 2,160 degrees Fahrenheit (1,200 degrees Celsius) the gas, the ore, and the flux all interacted. The flux combined ("flowed together") with the non-copper components of the ore to become slag. The copper component of the ore turned molten and filtered down through the lumps of porous slag to collect in a pool at the bottom of the smelter. When the furnace grew cool the coppersmith either dissassembled it entirely or removed only enough of it to retrieve the now-solid copper. Whichever he did, he left the slag behind; this smelter waste is what allows archaeologists to identify early metal-working sites.

Although copper is soft, it can be hardened by hammering, and so the first coppersmiths were able to make metal tools and weapons that had certain advantages over tools and weapons made only of stone and wood. Arrowheads, for example, could be cast in copper so that each one was exactly like the next; this uniformity is hard to achieve when working in stone. The same would be true of spearheads. As an added incentive for turning to metal weaponry, the uniform weights of copper arrowheads and spearheads would mean uniform performance in the field: given the same pull and the same cast (the user's responsibility), each successive arrow and spear should arrive close to the same place with close to the same potential for shock and penetration.

Early smelter was probably like this: a half-buried furnace made of fieldstones and left open at the top. One or two openings at ground level allowed air to be pumped inside. Filled at the bottom with charcoal and the rest of the way with ore and charcoal mixed together, the furnace is then fired. When the charcoal has turned to ash, if copper is being smelted, a puddle of metal will have collected at the bottom of the furnace. Here, iron ore is being smelted; at the

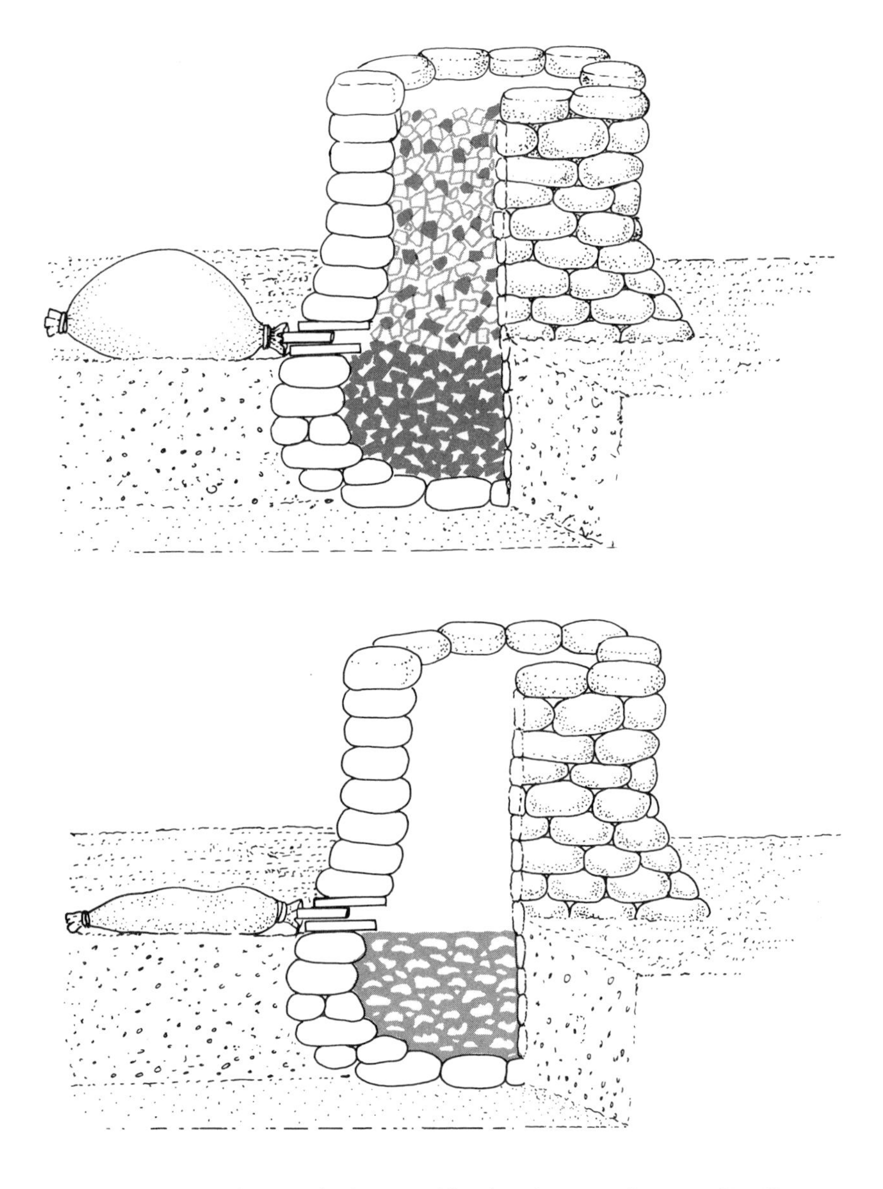

temperature of a simple furnace like this the iron does not liquify. It forms a spongy mixture of metal and waste for a smith to work on.

We are offered the hope that one day we need not study war any more. Hasten the day. Here at the dawn of the Age of Metals it is interesting to find that sword and plowshare are already about equally important as far as energy requirements go. Before the coppersmith arrived swords — as we think of them — were not known. The shock weapons, other than spear, dart, and arrow, were the stone battle-ax and mace. In other parts of the world, thousands of years later, a sword-like weapon was invented independently. This was a strong length of wood, scraped down until it was flat in cross section. Flint or obsidian blades were then set along the narrow edge with adhesive such as plant resin. In the Pacific, where suitable stone was rare, the weaponmakers ingeniously substituted sharks' teeth for stone blades. The coppersmith, however, having already cast replicas of the elegant stone daggers that the Neolithic knappers knew how to make, soon realized that a copper sword — in effect, a stretched dagger — was a superior weapon.

Copper tools were really more important than copper weapons. For example, you can make a first-class drawknife from unbending flint or obsidian and even fashion a fair stone chisel (the root of this word is the Latin *caesus,* from the verb *caedere,* "to cut"), but you need metal, that can be bent to the desired shape, to produce a really efficient gouge. Moreover, a metal cutting edge can be resharpened again and again.

Far more important than that kind of carpenters' tool, however, is the one that my phantom innovator evoked: the saw. This is perhaps the least appreciated of metal tools, yet no artifact made from stone, from wood, from leather, or from cordage can do what saws can do. Think about it. Until little more than two centuries ago, wood, stone, and fired clay were civilized man's major building materials. Little more than four centuries ago wood was also the principal material machines were made of. Does it seem odd that, some six millennia after the Age of Metals began, mankind was still so dependent on wood?

It should not seem odd. Wood, which our Mesolithic and

Neolithic forebears used as a raw material of tremendous promise, is marvelously adaptable to man's energy-hungry demands. Adaptable, that is, once an effective tool for shaping the tough, resilient, fibrous raw material becomes available. This tool is the saw.

Consider the carpenter's problem before the Age of Metals. If he has the appropriate straight-grained timber available to him (as, for example, the Indians of northwestern North America did), he can fell and trim a tree with a stone ax, let the log "season" for a year or two, and then split it into thin planks by using wedges and hammers. He can even square off the ends of the planks with a sickle-like kind of Stone Age saw: a grooved length of bone or wood in which sharp stone teeth are set. Okay?

Not really. Such a Stone Age saw can cut no deeper than the length of its teeth. Our proto-carpenter could not saw his seasoned log in half. He could not turn a non-splitting log into planks. He could not make any kind of deep saw cut because pretty soon the saw handle, three or four times wider than the saw teeth, would reach the surface of the log and the saw teeth could penetrate no deeper.

A metal saw makes Mr. Carpenter's troubles disappear. His friend, Smith, can pound a length of copper uniformly flat and thin, and hammer-harden it in the process. Saw teeth can be

"Greatest thing since the invention of the wheel," a common phrase, needs to be rewritten. Wheels (and their variations, such as gears) are certainly important, but these dingy bits of strip metal are what made wheels, gears, and all the rest possible. This is a Bronze Age saw from Minoan Crete; even earlier (and certainly even cruder) copper saws were the key to man's mastery of woodworking. The making of saws led with such speed to the making of wheels that no one knows if potters' wheels or vehicular wheels came first.

cut into one edge of the flat oblong and the saw blade will be no thicker than the teeth are. There is no limit to the depth such a saw can cut. If he wishes, Carpenter can now slice successive end pieces off a rounded log, thereby producing the first known disks of wood. As he will certainly want to do, he is now free to use other than straight-grained timber to make planks.

With metal saws, gouges, planes, chisels and drills, Carpenter's tool kit really lacks nothing it would contain today except a lathe (rotary motion) and those exotic engines that have converted ancient saws and drills into modern effortless rotary-motion gadgets. For that matter, if, early in the Age of Metals, Carpenter actually does slice off some log end pieces and manages to think about them constructively, he may be the man who ushers in the universal rotary-motion device: the wheel. He even has a choice between two kinds of wheel (they appear together at about this time): cartwheel or potter's wheel. I like to think that because Carpenter probably knew Potter far better than Packer, with his string of donkeys (that upstart, Carter, isn't even around yet, since wheels come before carts), he sent his first disks across to Potter for use as turntables.

Mr. Smith, meanwhile, was making many copper tools other than saws. He was casting axes for Mr. Forester and hoe blades and shovel blades for Mr. Farmer. He even hammered out a saw with no teeth for the Mason brothers, who used it at their quarry to square up their blocks of marble with sand and water as the cutting agent. And with each passing decade, looking at southwestern Asia as a whole, the members of the coppersmith clan were unconsciously and, later, consciously converting themselves into bronzesmiths, with the result that the tools and weapons they cast and hammered in their smithies were steadily improving in quality.

Before looking at the new methods of harnessing energy made possible by the Age of Metals we should see just why bronze is worthy of being called a great alloy. At the same time we will behold a wonder still not entirely explained after more than 3,000 years: the eclipse of bronze by man's second great alloy: steel.

Now, both metallic iron — from meteorites — and various iron ores were known during that subdivision of the Age of Metals scholars have dubbed the Bronze Age (3500 to 1200 B.C.). For example, Bronze Age smiths often turned iron into ornamental objects. One of the treasures from the tomb of Tutankhamen (1350 B.C.) is such an object — a dagger with an iron blade. The dagger's ornate handle, made of ivory, rock crystal, and gold, suggests that the weapon was ceremonial, rather than for day-to-day use.

The reason iron ores, which are plentiful, were not being smelted during the Bronze Age in as large quantities as copper ores is ridiculously simple. Cast copper and cast bronze are both useful metals. The copper is soft but it can be work-hardened,

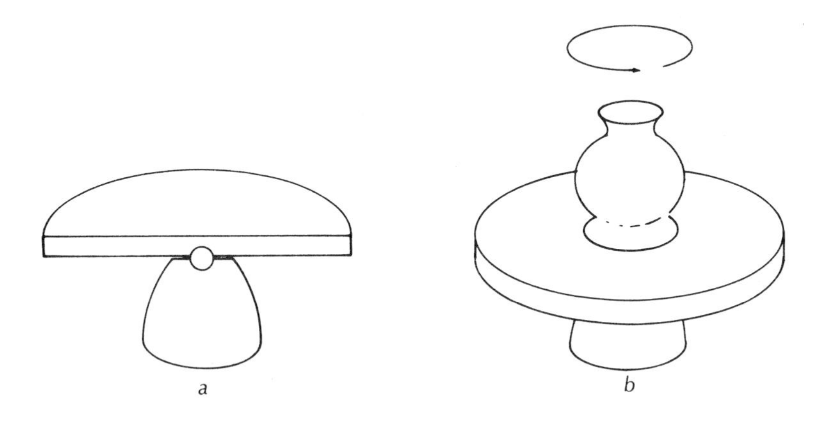

True rotary motion: the potter's wooden wheel, resting on a pebble bearing (*a*), can rotate continuously in the same direction. The rotation allows the potter to model the clay into a symmetrical container (*b*) swiftly, rather than building the pot with a series of clay "snakes." No one is sure whether the potter's wheel or the vehicular wheel came first, or were simultaneous inventions. It is clear, however, that pottery manufacture became a man's mass-production activity after the potter's wheel came into use early in the Age of Metals.

while the bronze can practically be picked up out of the mold and put to use. Bronze Age smiths, however, could not cast iron at all. The kiln temperature needed to reduce iron ore to molten metal is about 50 percent higher than the temperature needed to obtain molten copper or bronze. As a matter of fact, the smiths were not missing much; cast iron is so brittle that it is fit only for ornamental use.

Considering all this, you may well ask how it happened that iron ever came to be used at all. The answer is that, perhaps by trial and error but probably by accident, Bronze Age smiths learned that iron ore could be turned into a spongy mass (called bloom iron) at about the same lower kiln temperature needed to produce molten copper. This ugly bloom, a mixture of slag and sponge iron, when taken from kiln to forge, could, by persistent hammering at red-hot temperature, be cleared of the slag. The relatively pure iron that emerged from this forging process was not brittle like cast iron. When reheated it could be hammered again and formed into a variety of shapes, both ornamental and useful.

Forged iron ornaments outnumbered useful iron artifacts during the Bronze Age for three very good reasons. First, a smith could cast 20 bronze swords in the time it would take to hammer out only one sword made of iron. Second, a bronze sword is stronger than one made of bloom iron. For example, the tensile strength (that is, the resistance to stress of the taffy-pull kind) of pure cast copper is such that a load of less than 32,000 pounds per square inch (psi in abbreviation) will not stretch the metal. Bloom iron is only slightly stronger; 40,000 psi. The tensile strength of cast bronze, however, is 60,000 psi, and, when the bronze is work-hardened by hammering, its tensile strength can reach 120,000 psi. Work-hardening will only raise the strength of bloom iron to around 100,000 psi. The third and last strike against iron is that it rusts rapidly, whereas bronze corrodes very slowly.

With these three handicaps, particularly the third, it is not surprising that the very many Bronze Age sites in southwestern Asia that have been found to contain almost innumerable bronze artifacts have yielded fewer than 500 iron ones, most of them ornaments. To be sure, perhaps an equal number have vanished into rust, but even if ten times that many had vanished in this way the total number remains trivial compared to the thousands upon thousands of bronze objects that still survive.

At sometime late in the second millennium B.C. all of this changed for reasons that are still disputed and are certainly unclear. In the eastern Mediterranean the scene was one of military upheaval — the pertinent Egyptian texts speak darkly of the "peoples of the sea" — that extinguished a number of petty states, including those of Mycenaean Greece. The Egyptians themselves were hit twice within 50 years (in 1225 and 1183 B.C.). At Pylos in Greece, where the day-to-day palace records, written on clay tablets, were baked (and thus preserved) by a destructive fire the seaborne invaders may have set, the texts speak of a frenzied collection of all possible scrap bronze in order to make ". . . points for spears and javelins." The edict suggests a general shortage of new bronze.

Some analysts see this and other evidence as an indication that the bronzesmiths of southwestern Asia were finding tin hard to get at this time. Truth to tell, however, it is still not clear where any of the tin used to make all the bronze of the Mediterranean Bronze Age came from, much less why a shortage of tin should have developed just at the end of the second millennium.

Nevertheless, the evidence for a decline in bronzesmithing is impressive. For example, one English scholar has tabulated the number of knives, swords, and spear points found at Greek sites dating between 1050 and 900 B.C., that is, right at the turn of the millennium. Of the 15-odd knives, none was bronze; of 20-odd swords, one was bronze; of 40-odd spear points, eight were bronze. All the other knives, swords, and spear points were made of bloom iron. In Greece, at least, the Bronze Age had ended and the Iron Age begun.

All right; down goes the first of man's great metals. Ironically enough, the bronze we still make today is used largely for ornamental purposes, just as iron was used during the Bronze Age. All right, again; the expanded use of iron as a material for weapons coincides with a period in ancient Mediterranean history when bronze was apparently in short supply. The shortage, in turn, may have been because war conditions upset the tin trade. But what about afterwards? Things settled down eventually and so, at least in theory, the tin trade should have returned to normal. But bronze doesn't recapture the market. And this despite the three facts we have already noted; the fact that hammering out iron artifacts is much harder work than casting bronze ones; the fact that iron is not as tough as bronze; and the fact that iron rusts rapidly!

How can this be? A large part of the answer is that we have used the wrong word in the preceding sentence. Instead of writing "iron" we should have written "steel." Today's metallurgists think of steel as being essentially a modern invention with a trivial but amusing antiquarian history that goes back to the armor and weapons of the Middle Ages and even to certain smiths of Damascus early in the Christian era. The fact is, however, that the ironsmiths at the end of the second millennium B.C. were turning their soft bloom iron into hard steel as they repeatedly heated and hammered it. The process, known today as "steeling," was surely accidental at first. Technically, steeling occurs because the charcoal fire in the smith's forge is a natural source of carbon — the element that, when added in small amounts to iron, produces the alloy of iron called steel.

We know of a candidate for the title of first coppersmith. We know of none to contend for first goldsmith, silversmith, or bronzesmith, and certainly none for first blacksmith. Nevertheless, we can reconstruct almost exactly how he steeled his spongy lump of bloom iron. For every hour that the bloom iron was exposed to the fire in his forge it absorbed more and more carbon from the charcoal. In 20 hours of exposure to the forge fire, during successive reheatings and hammerings, the amount of carbon the iron will absorb rises to as much as 1 percent by weight. The bloom iron will not have become steel through and through because the outer surface of the bloom has absorbed most of the carbon. Even after 20 hours on the forge the steeled portion of the billet will not be much more than one tenth of an inch deep on each side. Whether our blacksmith has hammered out a sword blade or an axhead, however, he has produced an artifact with a tensile strength approaching 80,000 psi, which is substantially more than the strength of a bronze alloy containing 8 percent tin.

The first users of accidentally steeled iron weapons and tools must have reacted favorably to this tougher metal that would hold an edge longer than its bronze counterpart. The first blacksmith and his growing company of fellows, in their turn, would soon have advanced beyond the level of trial-and-error steeling. Still, there must have been more working for them than a shortage of bronze and the fact that they had overcome one of the three drawbacks of iron: its inherent softness.

Stainless (that is, rust-proof) steel is a product of our century, so the first blacksmiths certainly did not overcome the metal's third drawback. This leaves only the comparative work investment to consider. It is just here that the blacksmith had one major economic advantage: iron is incomparably less costly than bronze. Whereas copper is quite plentiful and tin reasonably so, iron is ridiculously abundant.

One fifth of the earth's crust is iron. Only three elements are more bountifully supplied to man; oxygen, silicon, and aluminum. The principal ores of iron — hematite, that same red ocher men have used as a pigment since the Old Stone Age, magnetite, siderite, and limonite, a variety of hematite — are readily recognized. In fact, the copper smelters who supplied the earliest coppersmiths with their ingots actually used hematite as the flux in their smelting process. (They also may well have produced the first sponge of bloom iron, just by messing around.) In any event, to produce a ton of bronze would certainly have cost twentyfold the cost of producing a ton of bloom iron. Consider the cost of the man-hours invested in moving tin ore or ingots, and possibly copper ingots as well, to the coppersmith's furnace.

This twentyfold cost advantage of iron over bronze is

largely canceled, however, by what happens next. As we have seen, the bronzesmith can cast 20 swords and still invest no more man-hours than the blacksmith puts into steeling a single iron sword. But the cancellation is not exact. The blacksmith still retains a substantial edge over the bronzesmith so far as investment in materials is concerned. Moreover, he has the edge of offering a superior product. And yet the bronzesmith has 20 swords. . . . Too many intangibles, including the value to be set on labor, enter into calculations of this kind. Nevertheless, the archaeological record speaks clearly: bronze out, steeled iron in.

If we cannot identify the first blacksmith, students of early metallurgy have presented us with a candidate for the first Class A blacksmith-craftsman. His workshop was in a small seaport, a Greek trading post on the Anatolian coast of Turkey, and he lived during the fourth century before Christ. One day he made an adz blade by hammer-welding two sheets of bloom iron while both were red hot, a common enough practice. But did he use two sheets of steeled iron? He did not; a woodworking tool need not be that strong. Instead he used a steeled-iron sheet for the outer surface of the adz and a soft-iron sheet for the inner surface, achieving a considerable saving in man-hours thereby. As is pointed out by the scholars at the University of Pennsylvania who have analyzed the smith's actions, when a blacksmith learned how to control the process of ironworking in such a subtle manner as this, mankind had passed a significant milestone in his mastery of technology. All honor to the forger of the bimetallic adz.

Sometime back we watched a bow drill in use. More recently we observed how a saw made of metal could slice end pieces off a log and how such little "wheels" may have inspired the potter's turntable. Now it is time to consider rotary motion in general, wheels in particular, and the application of power other than human and animal muscle-power to the Greeks' five simple machines: lever, wedge, endless screw, wheel-and-axle, and pulley.

As a starting point we can remember that the bow drill, like the hand-rolled fire drill and the many other kinds of drills before it, allows us to apply rotary motion only in a partial manner: the drill turns only so far in one direction before it must stop and begin to turn in the opposite direction. As that pioneer anthropologist, Gordon Childe, wrote a quarter of a century ago, "true rotary motion" means the freedom to turn in the same direction indefinitely.

One such device may be almost as old as the bow drill, assuming that the semi-settled people of Mesolithic times were using their wits at the same time they were making thread for their needles and string for their nets. This is an artifact almost unknown among civilized people today because all of us buy our thread on spools and our yarn by the skein. It is the spindle: a simple weight that includes a hole (or hook) so that it may be attached to a strand of thread, string, or yarn. We may call it the ultimate ancestor of the flywheel, although there is nothing wheel-like about it.

Thread, even the finest imaginable, is not a single fiber but a bundle of individual fibers that have been twisted together in an overlapping manner so that where one fiber ends another has already taken its place. Obviously, to keep the thread from unraveling, all the twisting must be in the same direction. The thread spinner, therefore, holds a loose mass of fibers in one hand, teases out a short, loose bundle from the mass, and attaches the spindle to the end of this loose bundle. Next, the thread spinner whirls the spindle and at the same time teases more and more fibers from the loose mass. (Unless you want lumpy thread, this takes very keen coordination, bringing forth just the right number of fibers at the right rate to be uniformly twisted into thread by the rotation of the spindle.) When the spindle begins to exhaust the momentum supplied to it by the thread spinner's whirl, the process is halted until the hanging weight is given another good whirl. Within the limits set by the strength of the thread, the heavier the spindle is, the fewer halts are necessary in the thread-spinning process. If we are charitable about the word "indefinitely" in Childe's definition of true rotary motion, surely the spindle can share billing with the wheel. Childe thought so, anyway.

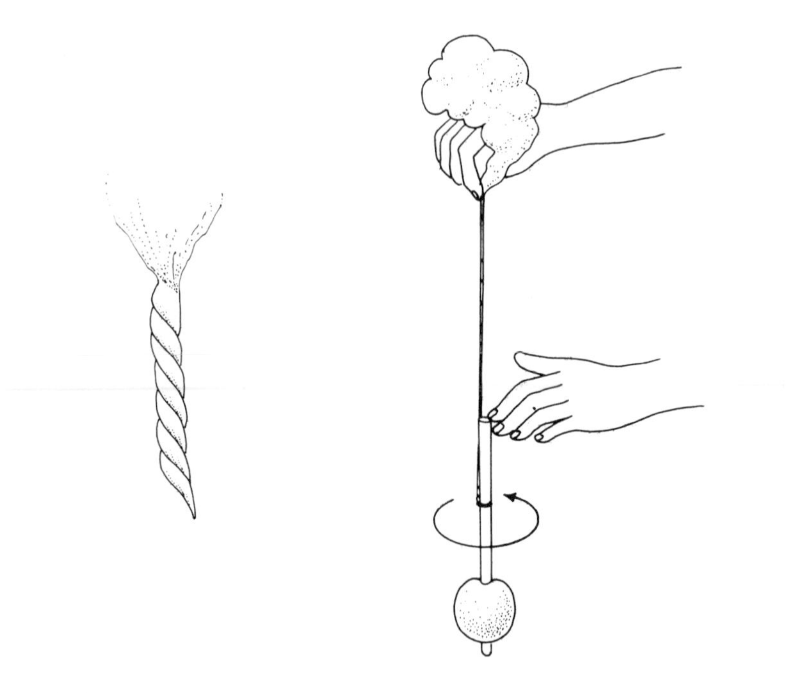

A simple rotating weight (here a lump of clay on a stick), the spindle may be the ultimate ancestor of the flywheel. As it turns, it twists individual plant fibers into thread. Like the bow drill, the spindle was one of man's first tools to utilize the principle of rotary motion.

This simplest of machines — a mere lump of clay — was certainly of great technical importance to spinners of all description, from net knotters to knitters and the most advanced group of all, clothweavers. In spite of its practicality, however, I am tempted to say that even more important is the flywheel principle the spindle incorporates. As an energy-storage device, the spindle is exactly analogous to the spring: one brief investment of muscle-power — winding up the spring or whirling the spindle — is subsequently spent in a less energetic fashion over an extended period of time. In this connection, one peculiar version of the bow drill, known as a pump drill, actually incorporates a flywheel.

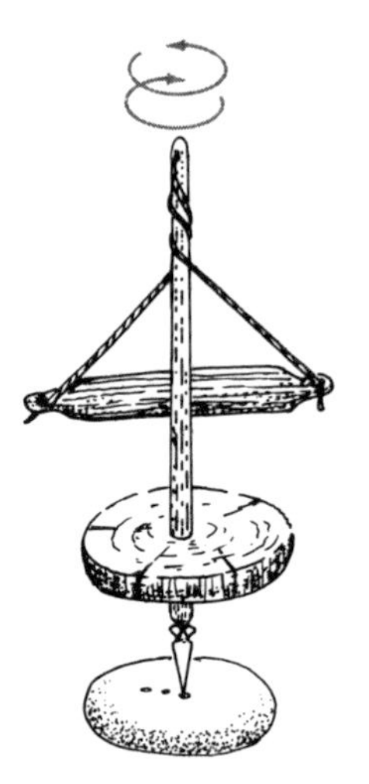

Not quite true rotary motion: like its simpler forerunner, the bow drill, the pump drill (this one equipped with a flywheel), turns first in one direction and then in the other. It was a very good tool for making holes, once fitted, as here, with a long-wearing metal bit.

To pause for just one moment, let me mention three other bow-drill variations and thereafter put behind us these devices for supplying interrupted rotary motion. One of these is the core drill. Instead of having a solid cutting bit at its business end, this drill was equipped with a hollow, cylindrical bit. In the days before metal, hollow reeds were probably used as bits. The hafting holes that were cut through various Neolithic stone axes could have been made with reed core drills, using an abrasive such as sand to do the actual cutting. This must have been extremely tedious as well as delicate work. Once even so relatively indifferent a metal as copper became available, the core drill came into its own and all kinds of holes could be made in stone at only a fraction of the power investment required for non-core hole-making.

Drilling on a grand scale: these two Egyptian craftsmen are hollowing out a soapstone basin and urn with drills weighted at the top. Although this is the Age of Metals, their drill bits were probably flint. Note that both the vessels appear to have been shaped on a lathe.

A second bow-drill variation involves a difference in degree rather than in kind. This is boring, which is to say, drilling on a grand scale. In talking about containers and the work of the first potters I mentioned that vessels made from soft stone, such as soapstone, were among the earliest inorganic (that is, not made of skin or plant materials) containers. In Egypt the production of stone vessels, hollowed out with boring bits made of flint, had become nearly a belt-line business by the third millennium B.C. Here again the flint bit did scarcely any of the cutting directly: sand or emery was the abrasive.

Finally, what do you have if you devise a drill bit that is rigidly fixed in place and then rotate the object you wish to shape — for example, the outside of a stone vessel? You have the rotary-motion device we call a lathe. In effect, this is a bow drill turned on its side, with the rotary motion of the drill device being applied to the work rather than to the cutting head. Again, the rotary motion is not continuous. but being able to rotate the work at considerable speed for even a short time is a great advantage. This kind of machine appears to have a respectable — but unproved — claim for an early origin in southwestern Asia. In a form called a pole lathe, with a springy timber taking the place of a bow, it continued in use in Europe and the Mediterranean basin until long after the general availability of continuous rotary motion. End of pause.

Again, not quite true rotary motion. The piece of work on the lathe turns first one way and then the other as the lather pushes the pedal down with his foot and the springy pole then pulls the pedal back up again. The addition of a crank to the lathe would permit true rotary motion but would require another man to turn it.

Which came first, vehicular wheel or potter's wheel? Argument seems fruitless. It is certainly possible that the artfully weighted turntables, with their stabilized axle supports and smooth-running bearings that revolutionized Mr. Potter's art, were independent inventions, owing nothing to Mr. Carpenter's new-won ability to detach circular sections of wood from a tree trunk. For that matter, it is by no means absolutely sure that the earliest solid vehicular wheels were the result of this particular sectioning trick. What seems crystal clear, however, is that the potter's wheel, as soon as it appears, transformed what was earlier a cottage craft and probably women's work into an industry conducted mainly by men.

The oldest potter's wheel known was found in the 1930s by the excavators of Ur, among the early levels of that early Mesopotamian city-state. It is thought to date to the late fourth millennium, say about 3250 B.C. There are surely still older potters' wheels awaiting discovery, but the wheel from Ur showed the two basic elements of this great invention: the combination of a flywheel and a lubricated pivot. With both a smooth bearing and enough weight to keep the turntable running after the potter gives it a whirl, a device like this will rotate continuously in the same direction at speeds greater than 100 revolutions per minute. The prepared clay the potter sets at the center of the spinning turntable can be raised up into the shape desired with only the lightest pressure from the potter's hand to guide it. To make such a "thrown" pot takes only about 10 minutes, compared to the average of 10 hours required to form the same pot by coiling. This is an increase in efficiency by a factor of 60 to one. No wonder that a former cottage craft suddenly evolved into a major manufacturing enterprise.

So here we say farewell to Mr. Potter, his forebears, and his descendants. He will achieve many more advances but all will be matters of degree and not of kind. He will learn to select and even to refine better clays and he will develop better ways of firing the pots he makes. His experiments with different minerals for painting and glazing his pots will in no small way provide the basis for chemistry as a field of investigation and his kiln improvements will open the road to more refined metallurgy. All the same, these steps, however significant, are all little steps. The big steps were made by others. By the mammoth hunters of Czechoslovakia, for example, who remain the first to have fired clay, late in the Ice Age. By the Neolithic makers of coiled clay containers, who were the first to exploit the physical alteration that comes when clay is exposed to heat in a kiln-like environment. And finally by Mr. C.C. (for "clay coiler") Potter's great-great-grandchildren, of the Copper or early Bronze Age — who were either given a fresh-sawn stump by Mr. Carpenter or invented the flywheel turntable all by themselves.

Count the years. From, say, 4000 B.C. to today amounts to a little fewer than six thousand of them. From (what shall we call him?) T. T. (for "turntable") Potter back to his ancestor C. C., the founder of the line, is a matter of some six thousand years more. Add 10,000 or more years to that total in order to reach back to the figurine-bakers of Vestoniĉe. Then the interval separating T.T. from today is quite brief — only about one third as long as that separating T.T. from his ultimate clay-player forebear of the Ice Age. Here again we seem to glimpse something of the so-called acceleration of history. Agreed? In any event, what C.C. and, later, T.T. did are the major events in mankind's happy involvement with clay. All that has happened between T.T.'s day and the rise of M.M. (for "modern manufacturing") Potter is trivial by comparison. Farewell, then, all you Potters with your speedy little wheels. What about the Big Wheel?

A generation or so ago the Big Wheel question was easy to answer. In that same Sumerian city-state, Uruk, where several of man's earliest written words resembled in outline the clay tokens used by Neolithic tradesmen, two words are unmistakable pictographs. One of these shows what looks like a heavy-duty travois with a square, roofed shelter at the rear end of the skids; if snow fell in southern Mesopotamia, this could be a drawing of a sled for sissies.

You may think that the interpretation of this pictograph is, if not farfetched, at least remarkably imaginative. And so might any man, were it not for its companion. The second pictograph shows an almost identical travois-and-shelter, although it is headed in the opposite direction and the front ends of the skids are more upcurled. Beneath the skids, however, are two small circles. This pictograph simply has to represent a four-wheeled cart, although I am sure that all the scholars who support this interpretation would sleep better at night if only a pictograph of a draft animal had been included (as was done by the Chinese when they represented a war chariot with spoked wheels some 1,500 years later).

So we have a wheeled vehicle represented in the written language of Uruk. We even have a suggestion of how this wheeled vehicle was invented: you put two axle beams underneath a heavy-duty sled and add a wheel to each end of each

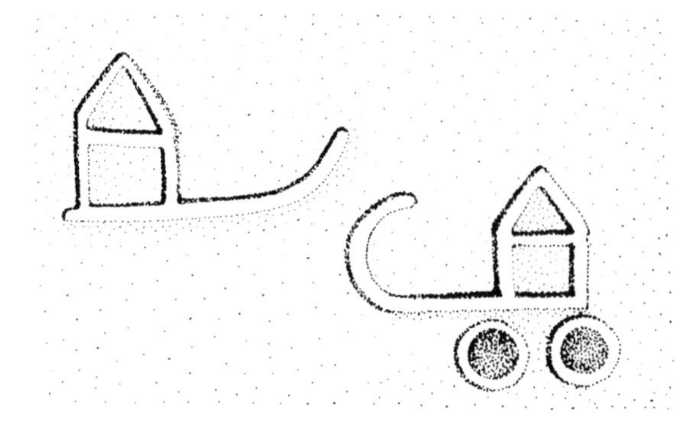

Earliest depiction of a wheeled vehicle, at right, was drawn about 5,000 years ago by a scribe at Uruk, a Sumerian city-state of early Bronze Age times. It combined the Sumerian pictogram for a sledge, at left, with pictograms representing a pair of (solid?) wheels.

axle. Because the development of writing at Uruk is dated to about the end of the fourth millennium B.C. many scholars logically concluded that the first wheeled vehicles were invented in southwestern Asia, and most probably in Mesopotamia, before the start of the third millennium B.C. The conclusion was more easily accepted because in the early centuries of the third millennium, in about 2700 B.C., the Royal Tombs at Ur and at Kish and even at not-so-distant Susa all include wheeled vehicles, both two-wheelers and four-wheelers. All are equipped with disk wheels; disks made from three or more planks are more common than one-piece disks.

This elegant example of invention, evolution, and diffusion, blending as it does the careful work of dirt archaeologists with the puzzle-solving abilities of the scholars who interpreted the inscriptions, fell apart soon after World War II. During the postwar years and on into the 1960s excavators working in three areas — the Caucasus, Soviet Armenia, and the Kalmyk Steppe (near the delta of the Volga River) — have unearthed abundant evidence that wheeled vehicles were first known there about as soon as they were recorded in the Uruk wheel pictograph and perhaps even sooner. In the centuries thereafter, these northern peoples showed great technical talent. For example, at the Kalmyk Steppe site of Elista a four-wheeler was found complete with a pivoting front axle. This is the earliest steerable wagon known; the next one in the archaeological record, the Dejbjerg wagon, dates to the final century before Christ, and steerable wagons were not in common use in the West before the Middle Ages.

As a second example of northern ingenuity, some of the wagons unearthed at Lake Sevan, a site in Armenia, had spoked wheels rather than disk wheels. The steerable Kalmyk wagon was built about 4,400 years ago; the Armenian spoked wheels were assembled more than 3,000 years ago.

Spoked wheels could easily have been invented twice. They turn up even earlier to the south and west of Armenia, always in two-wheeler applications associated with horses. This is the classic war chariot that continues virtually unchanged from ancient Egypt and Mesopotamia via the Mycenaean Greeks (and Homer's siege of Troy) down to Queen Boudicca's pitiable effort

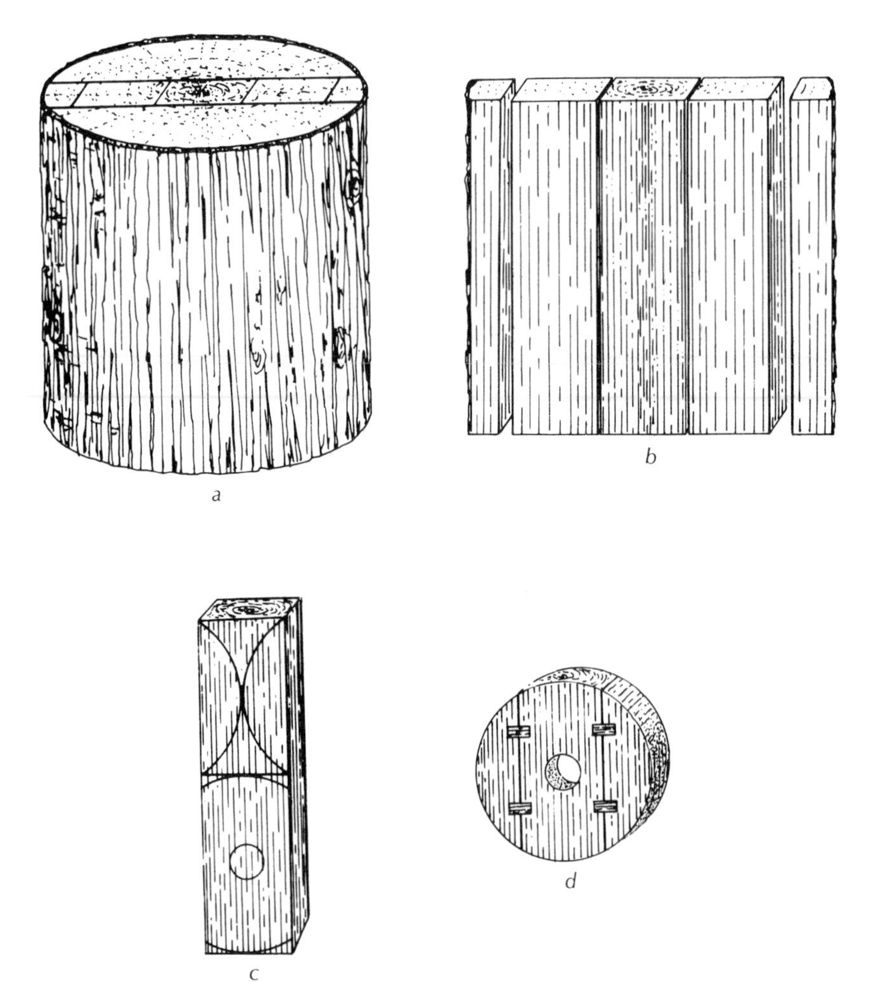

How to make a disk wheel the easy way. From a convenient bit of tree trunk (*a*), saw out a center section and cut it into five beams (*b*). Take the middle beam and scribe it (*c*) to guide further sawing and drilling. Then assemble the three pieces tongue-and-groove style (*d*).

to halt the Roman advances in Britain in the first century A.D. This also almost certainly means that men used horses as draft animals before any but a few steppe herders thought of riding astride a horse. In fact, serious riding astride — as exemplified, say, by squadrons of cavalry — could not have arisen much before the middle of the first millennium A.D., when the invention of the stirrup somewhere in civilized Asia first allowed the horseman a sporting chance of keeping his seat.

The appearance of the spoked wheel spelled the doom of the disk wheel that is by comparison hideously inefficient. Patient and well-muscled oxen will move a disk-wheeled wagon; after all, as we can surmise, they were dragging sleds overland before the first disk wheel appeared on the scene. When you harness up horses or mules, however, you want to hitch them to a livelier rig. Perhaps the last one-piece disk wheels made anywhere in the world are represented by one that was excavated in the Netherlands some 20 years ago. It is a piece of patient carpentry wonderful to behold, hewn laboriously from a vast plank. To

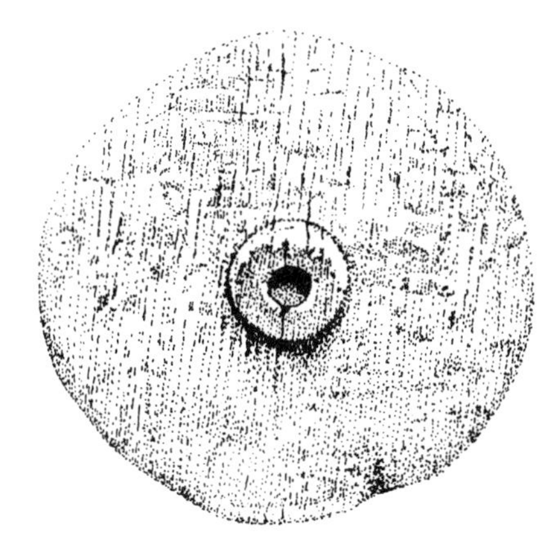

This one-piece wheel, laboriously sawn and adzed from a heavy plank originally at least three times thicker (look at its hub), was made by carpenters in the Netherlands some 4,000 years ago.

judge from the breadth of the hubcaps, the plank must originally have been three times the thickness of the finished wheel. It was made late in the third or early in the second millennium B.C.

A few three-piece to five-piece disk wheels are made to this day for use with ox-drawn two-wheel carts. These vehicles are the delight of government Tourist Bureau photographers around the world. When they can, however, the peasants (who actually use the carts) prefer to mount them on salvaged automobile wheels. Complete with pneumatic tires!

So much for vehicular wheels, then. Those with an interest in cause and effect — a delightful, if idle, pastime — will take note that no wheeled vehicles were independently invented in the New World. They may choose to correlate this fact with the absence of any draft animals among the domesticated animals of the pre-Columbian Americas. Maybe. I find it more puzzling that so many civilized New Worlders, familiar with rotary motion at the Mesolithic level of the bow drill and fire drill, kept making pottery the hard way up to and even beyond the time of contact with wheel-knowledgable Europeans. This is all the odder because the principle of continuous rotary motion had been recognized in at least two parts of the pre-Columbian New World: Peru and the Valley of Mexico.

In Mexico, although not much is made of it by archaeologists, the little pre-Columbian children were furnished with pull-toys just as our own toddlers are today. And just as today, the toys rolled along on wheels. In Peru, even more surprising, not only a potter's wheel but also a lathe was used to produce grave offerings that were unearthed recently at Pashash in the highlands north of Lima (the investigator was from the University of Texas). The grave is that of a woman, evidently of some note, who died in the latter half of the fifth century A.D. The wheel-turned pottery that was buried with her consisted of scores of pedestal cups; the cup and its foot were wheel-thrown separately and then joined.

The evidence for use of a lathe at Pashash is the presence in the same woman's grave of 15 other cups cut out of stone.

Evidently each stone blank was rigidly mounted on a shaft that was free to rotate; in all probability a flywheel was also mounted on the shaft to counteract the resistance of the blank to the cut of a fixed bit. Such a rig could be set in motion by wrapping a cord around the shaft and pulling hard, in the same way the disadvantaged among us start an outboard motor.

The Texas excavator, Terence Grieder, reports that the graves of other equally prestigious residents of Pashash, placed to rest some 200 years later, include no such products of rotary motion. Evidently this great technical leap forward, of vast importance in the Old World, appeared independently in the Andes, only to disappear again in a handful of years without having diffused any farther. Why? Try this proposal.

If, evidently by sheer coincidence, we find potter's wheel and vehiclular wheel arising at about the same time in the Old World, only part of the reason is because of the advance in carpentry (the saw) that the Age of Metals brought. Another part of the reason is that draft animals — plow-yokes of oxen — were already dragging sleds from place to place there. The yokes of oxen, in turn, were there because the principal grains — wheat, barley, millet, and even rice — grow best in plowed fields.

The principal foodstuffs of the New World — squashes, potatoes, manioc, and maize — are all well suited to hoe agriculture. A seasoned cause-and-effecter can imagine that some pre-Columbian prairie culture hero, recognizing the potential of plowland maize, might at least have harnessed a few volunteer squaws to a modified hoe and thereby independently invented the plow. Had that happened, the next step might easily have been domestication of the bison as a draft animal.

This did not, in fact, happen; the hoe continued triumphant until well after Cortez. With nothing to hitch to a scaled-up pull-toy, the people of Mexico were content to leave the wheel to child's play. The reason why a team of docile bison was absent, however, is far likelier to lie in the efficiency of hoe agriculture than in a lack of interest in animal transport. Once introduced by the Spanish, both the horse and the ox were enthusiastically and almost instantly adopted by the native population throughout the New World.

The Pashash problem is entirely different. A million overworked New World potters would have cheered themselves hoarse if the wheel-throwing potter of Pashash had come downhill to share her (or his?) invention with others. Yet this did not happen. Why? The only reply I can offer is one long since immortalized by the near-immortal Mr. Vonnegut: "So it goes."

CHAPTER
9

Water Power, Wind Power, Animal Power, and Man Power

LET THOSE splendid craftsmen, Potter and Carter, now withdraw to the wings. Smith and Carpenter may remain, because they will still play a substantial part in realizing the truly revolutionary aspects of rotary motion. This is the ability of rotary motion to convert machines into prime movers. All that is required to achieve this is the modification of wheels, and wheels will now be modified by the score and the myriad.

Here comes the water wheel: possibly the earliest application of non-muscular energy to the task of providing rotary motion. Right behind this "mill wheel" comes the millstone: a great heavy wheel made of a material sturdier than wood and installed lying flat. Between mill wheel and millstone lies a power train of gear wheels and cog wheels — now all made of wood but foreshadowing an untold number of ingenious metal cogs and gears, enormous, tiny, and every size in between. Windmills: using another kind of non-muscular energy, they also provide a source of rotary motion. A windmill's sails are the specialized spokes of a modified rimless wheel that, like the water wheel, drives a power train of other modified wheels. The list is virtually endless and literally staggering. It will be easier to bear if we approach it bit by bit, considering first of all how the Greeks' five simple machines relate to various kinds of prime movers.

The five, to repeat Hero's words, are those machines "by the use of which a given weight is moved by a given force." The Greeks listed them, in order of complexity, as the lever, the wedge, the endless screw, the wheel and axle, and the pulley. Lever and wedge we have already met in such simple guises as digging stick and hand ax; we have met the lever a second time in the more complex form of the spear thrower. The Greeks, I am sure, thought of the wedge in a more restricted sense than that of a cutting or splitting device: for them, wedges were the means whereby puny men could raise weighty masses above ground level. Wedges were, in effect, solidified increments of lever action. To give an architectural example, one can use wedges to "jack up" (as we say today) a great block of stone until rollers may be set in place beneath it, allowing the block to be rolled into the desired position. Many men with levers could do the same thing, but not so easily or safely. Once in place, the block can be wedged up again so that the rollers may be withdrawn; when the wedges are then knocked out, the stone is permanently set in place.

Now, levers and wedges do not require the application of force in the form of rotary motion. Neither does the pulley. Although pulleys incorporate small wheels, the force that sets them to work is linear — a long haul on a rope. For the endless screw (which we will examine later) and the wheel and axle, however, force can only be applied in the form of rotary motion. And, as Hero declared, the wheel and axle, like the other four simple machines, makes use of the same single principle. Consider our first example of a prime mover, the water wheel. Where the application of force is concerned, we can look on it as a spoked wheel: a wheel with many spokes, and every spoke a lever.

Whose was the first water wheel? The answer may never be known because wood is all too perishable. In speculating about this particular "first," however, it will be useful to consider that water wheels can serve two quite different purposes. The one we think of first, to the almost total exclusion of the other, is use of the water wheel as a prime mover: the force of the water

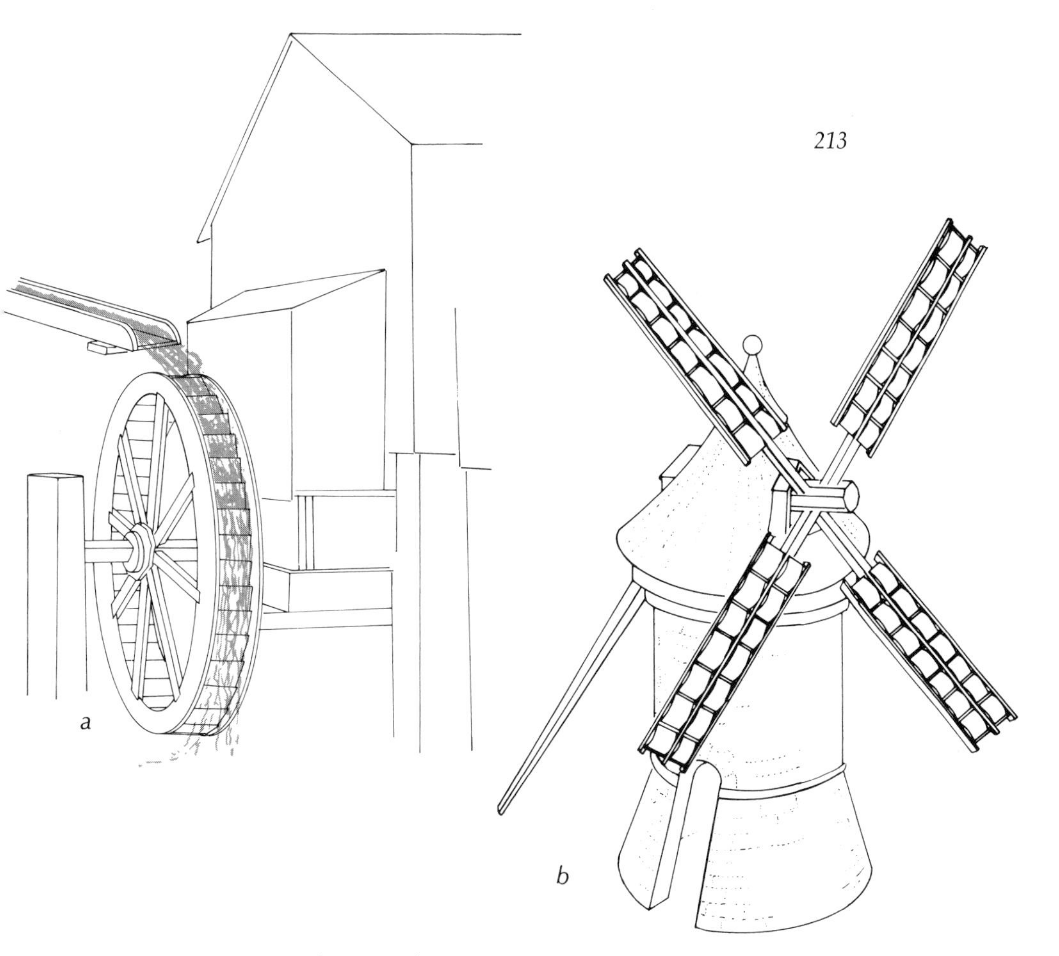

Water power, wind power: here are the two kinds of "wheels" that let man break away from dependence on his own muscle or the muscles of animals in applying energy. The water wheel (*a*), at left, actually looks like a big wheel with buckets around its rim (the one shown here is an "overshot" model; it receives the water from above). Water wheels were probably man's first prime movers. The rotary motion they supplied to their axles could be used to drive all kinds of machinery, either directly or through a train of gears. The motion was almost certainly first applied to millstones that ground grain into flour. The windmill (*b*), at right, looks at first glance less like a wheel until one realizes that it is a wheel without a rim that has wind vanes for spokes. This, the second of man's prime movers, was also first used to grind grain. A non-rotary application of wind power — the sail — was certainly used long before either water wheels or windmills were invented.

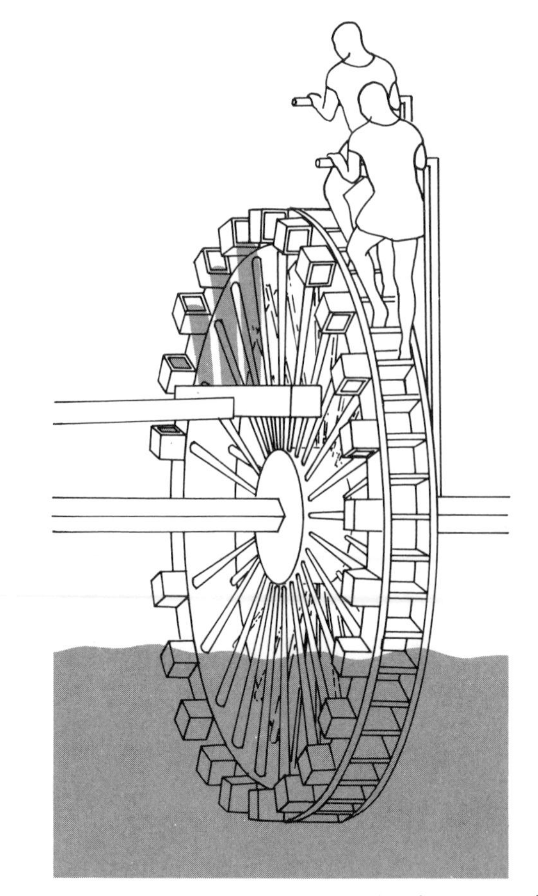

Possibly the earliest form of a water wheel: a water lifter. Wheels like this one, driven treadmill-fashion by men, were used to keep mine pits dry before efficient pumps were made. Note that in principle this wheel and all others like it are groups of levers.

makes the wheel turn and the rotating axle does the work. The second purpose is surely as old as (and possibly much older than) the first: the use of the water wheel as a water lifter.

In the water-lifter role, the prime mover is not the wheel but an animal, a man or several men whose muscle power makes the wheel turn. Where each spoke reaches the rim of the wheel the builder has constructed a shallow scoop that will hold water. The water load is picked up as the scoop passes up from its bottom position. The load is dumped into a trough as the scoop approaches its top position. How high each scoop of water is lifted depends on the size of the wheel.

This is, of course, another matter of lever action: each wheel spoke is a lever, with the axle serving as a fulcrum (the scholar's word for "prop"). If the lifter is man-powered, the human prime mover would probably "climb" one side of the wheel, using his own weight to make the wheel turn. If it is animal-powered, the horse — or donkey or ox — would probably tread a circular path, turning a horizontal wheel that would be geared to the axle of the vertical lifting wheel. In either case, the water lifter would, in fact, be little different in principle from the early Neolithic irrigation sweep, which is a simple counterbalanced lever-scoop combination.

A one-spoke water "wheel": this hinged lever raises the water from the well (or a stream, for that matter) by first sinking the bucket at its business end and then raising the counterweighted bucket to a trough. After that, gravity moves the water to the reservoir.

Was the water lifter the first water wheel? Only indirect evidence suggests this. When that marvelous Roman know-it-all, Marcus Vitruvius Pollio, approached the end of his 10-volume treatise on architecture around the start of the Christian Era (he was writing for one of the early Caesars, perhaps Augustus but certainly none later than Nero), he devoted a few chapters to water engines. In his view, at least, devices for lifting water received first priority; he dealt with them in Chapter IV and the first part of Chapter V and was rather offhand, in the latter part of V, in discussing a cog-wheel modification that permits a stream-driven water lifter to be converted into a flour mill (the famous Vitruvian mill).

This could represent nothing more significant than bias on the part of the author. Nevertheless, the question of which came first, waterlifter or water mill, just might be answered one of these days. This is possible because water lifters were in use, to keep mines dry, at least as long ago as the centuries immediately preceding the Christian Era. Who knows when the timbers of some far more ancient water lifter, sheltered from decay deep inside an abandoned mine, may be discovered and the age of the wood determined by carbon-14 analysis? Just don't hold your breath until this happens.

At present the oldest known water lifters are those once used in a copper-rich part of Spain, Rio Tinto. The mines there may have been first exploited by the Carthaginians. In any event, the Romans later worked the Rio Tinto mines for hundreds of years, beginning in the second century B.C.

In one of the mines the water lifter consisted of eight sets of wheels, each 15 feet in diameter. Allowing for the slope of the troughs that fed the water from wheel to wheel, the vertical lift per wheel was a little less than 12 feet. This means a total lift of slightly more than 90 feet. This is nothing to sneeze at; the greatest lift possible with a suction pump is scarcely one third as high.

R. J. Forbes of the University of Amsterdam, a first-rank historian of technology, finds the first written reference to a prime-mover water wheel (used to turn a millstone) in the verses of a Greek poet of the first century B.C. The first non-literary reference Forbes finds in the works of the Greek geographer Strabo. Writing long after the event, Strabo states that, when the great rebel-king of Pontus, Mithradates, built a new palace in the closing years of his reign, he equipped it with a water wheel for grinding grain; this would have been in about 65 B.C.

Strabo makes no great fuss about the king's action and certainly advances no claim that Mithradates invented the device. So it would appear that the water wheels of the eastern Mediterranean are about as old as the water lifters of the western Mediterranean and there the matter stands for now. Certainly, whoever saw either kind of wheel at work would have grasped the usefulness of the other kind on the spot.

Both the eastern and western Mediterranean wheels were set vertically, with horizontal axles. The poet quoted by Forbes, Antipater of Thessalonica, speaks of the stream of water ("Demeter's nymphs") "leaping down on the top" of the wheel. This vertical feature is more important than it may seem at first glance. Think for a moment. What is so wrong with a horizontal wheel on a vertical axle?

If anything, a horizontal water wheel seems more practical. The upper of the two millstones (the one that turns and grinds the grain) can be hitched directly to the rotating axle of a horizontal wheel. Thus there is no need to build a power train to carry the wheel's rotary motion through a right-angle turn. Must this not mean a net saving in energy, even though it is only Mr. Carpenter's energy that is saved by his not having to build the power train?

The answer to this question is a resounding "no," and that is why horizontal gristmills did not flourish. But I see that, once again, you are ahead of me. That's right: with a horizontal mill, one turn of the water wheel means one turn of the millstone; this is a direct drive. But the carpenter who constructs the power train for a vertical water wheel would need to be a lackwit to accept a one-for-one ratio of rotation. Let's take a look at his options.

Using the Rio Tinto wheel (15 ft. diameter) as a model, consider the following system. The drive wheel — that is, the wheel attached to the axle of the upper, rotating millstone — is made three feet in diameter and is fitted with 30 peg teeth evenly spaced around its circumference. To achieve the necessary right-angle turn in the power train our carpenter must fit a cog wheel at the far end of the horizontal water wheel's axle. This he does in such a way that the cogs mesh with the peg teeth on the millstone axle wheel.

If our carpenter makes his cog wheel the same diameter as the drive wheel, so that both wheels have 30 teeth, he will end up with nothing better than the one-for-one ratio that a horizontal water wheel provides automatically. In fact, he will be even worse off, for his big vertical water wheel cannot turn as rapidly as a small horizontal water wheel and so his millstone will be turned less rapidly than one driven by a horizontal water wheel.

What our carpenter must do is to make his cogwheel larger than his drive wheel. For example, if it is large enough to be fitted with 60 cogs, every time it rotates once it will make the millstone rotate twice, and so on. Although these first power trains were made only of wood and required endless maintenance and tooth replacement, they represent man's first advantageous gear ratios. (The transmission in your car, although much more complex, does a similar job but backwards. That is, the gears are so arranged that the engine crankshaft rotates many times for every single turn of the car's drive wheels when in "low" gear and fewer times when in "high" gear. With the gristmill, of course, there was no shifting of gears.)

All this may seem too hypothetical to you, so I hasten to offer some concrete evidence. Forbes calculates that a millstone turned by a donkey had an output of about 16 pounds of flour per hour. With an animal as a prime mover, an operating speed in excess of two revolutions per minute is unlikely: Moreover, the heavier the millstone (and thus the greater the leverage required to turn it), the fewer would be the revolutions per minute. The Romans built one mill near Naples that was driven by a water wheel less than six feet in diameter. It was geared so that the millstone, by Forbes's calculation, made 46 revolutions per minute. He estimates its hourly output at 330 pounds of flour, roughly the equivalent of 20 donkey-mills.

Nor was this any Roman record setter. Near Arles, an important military supply base for the Roman garrisons in southern France, the authorities took advantage of the water available from an aqueduct to contruct a hillside cascade of water wheels. They numbered 16 in all, arranged in pairs, and each six-foot-diameter wheel drove one millstone. Working a ten-hour day, as this tandem array frequently did in the third century A.D., it delivered 28 tons of flour a day to the Roman commissary.

The first actual pictorial representation of a water wheel comes from 5th-century Constantinople, seat of the Eastern Roman Empire. It is a palace mosaic and shows an undershot wheel. This kind of wheel is less efficient than an overshot one because it gains nothing from the weight of the water. At the same time, any old stream will do to drive an undershot wheel, whereas one must construct a dam and millpond to supply water to an overshot wheel. Still, in the end, most water mills were equipped with overshot wheels.

From what has been said up to now, you might easily think that the harnessing of water power was an achievement devoted entirely to grinding grain. This is not so. Mr. Miller — or, more appropriately, a host of Miss Millers, working on their knees — may have been the first to benefit but, as we shall see, Mr. Fuller was not far behind, nor were Mr. Smith and Mr. Miner. Indeed, many Millers of the Christian Era soon left their millponds to these others and moved off to occupy various open spots where they harnessed air power instead.

Today, some experimental windmills are being used to generate electric power. Day before yesterday in the U.S. (and still today in many parts of the world) windmills were the water pumpers farmers depended on. Before that, windmills were mostly quaint antiques. From abut A.D. 1000 to A.D. 1800, however, the grain-grinding Millers were the primary land-based users of air power. At least in broad outline, if not in fine detail, the windmills' history is easy to trace. Easy, that is, if you steer clear of historians.

Contrary to popular opinion, the first man to cry "I want it in writing" was not a lawyer but an historian. Lacking a document, your historian is struck dumb, and not always only in the sense of becoming speechless. For example, if one asks the historian to identify the first windmill in England he will proudly point to the rent rolls of the Knights Templar (a powerful body of leftover Crusaders) which show that a windmill at Weedly in Yorkshire earned the Templars an annual eight shillings' rent in A.D. 1185. He will with equal pride go on to note, (as the great Lynn White, Jr.'s essay on the subject testifies abundantly) that certain references to earlier windmills both in England and in France are either fraudulent or invalid. Ask him further if the Yorkshire windmill was the first in all Europe and our friend the historian can only reply that it is the first certain one; that is, the first attested in an unchallenged document.

Alas for us, at least so far, help from the archaeologist is not forthcoming. If abandoned, how long could a simple windmill, built all of timber, survive the seasons? Once fallen in decay, might it be recognized for what it once was? The later, more elaborate windmills of the Middle Ages, part masonry in construction, might be recognizable. But these were probably not left to rot anyway, at least not until better ways of grinding grain

Two kinds of water wheels: at the top is an overshot wheel, of the kind already seen. The one at the bottom is known as an undershot wheel, because the water strikes it at the bottom. Note that the two wheels turn in opposite directions. A third kind of wheel, not shown, is called a breast wheel. The water strikes a breast wheel at about mid-point and the bucket arrangement around its rim is designed accordingly. Still a fourth kind of water wheel, also not shown, was a horizontal wheel; it had vanes rather than buckets for the water to strike and was the precursor of all turbine prime movers.

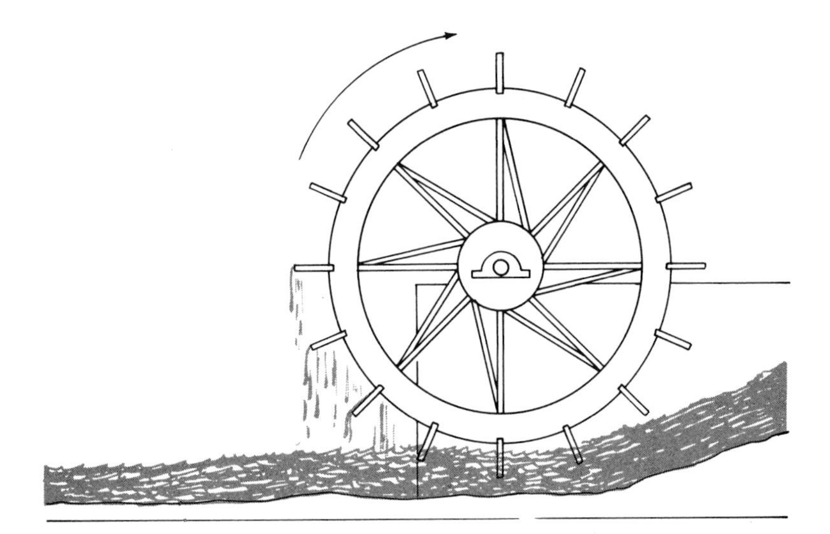

were devised. And so we are left with the conclusion — for lack of a better one — that the inhospitable eastern shore of a long-abandoned Roman frontier state, resettled by Germanic pirates, plagued by Norse ones, and recently overrun by some half-civilized, half-Viking hedge-knights from Normandy, was the site of a major advance in the history of technology.

Surely we deserve something better than that! And perhaps we have something better, although the small glimmers of evidence are not of the kind that meet the high standards of historians. In Persia, where the population has been plagued by extremes of summer heat and winter cold for many thousands of years, the first architects of the Age of Metals began to challenge the tyranny of the seasons first in simple ways and then in more complex ones. In essence, they made the air work for them.

The architects did this by means of a wind tower, in essence a chimney with no fireplace. On a still night, doors set at the bottom of the tower could be opened to allow the warm air inside the house to rise up the tower, pulling cool night air at ground level into the house. If there was a night wind, the direction of flow could be reversed, letting cool outside air come down the tower to force out the warm air inside the house.

These wind towers did more than make life comfortable for those folk wealthy enough to construct them. Together with another architectural feat, the underground aqueduct (primarily a Persian irrigation system that need not concern us here), the wind tower allowed the Persians to set up habitable outposts in their own desert fringes and in the inhospitable wilds of their neighbor states, particularly along the Arabian shore of the Persian Gulf.

When did all this happen? It would be unfair to say that no one has bothered to find out, but the fact remains that no one is sure. The dynasties that succeeded one another in Persia proper had little interest in preserving the humble administrative records of previous regimes. For example, shaft holes of many underground aqueducts appear like pockmarks on aerial photographs of the ruins of Persepolis, capital city of Persia in the first millennium B.C., but no attempt has been made to discover whether these irrigation systems were constructed before or after the great public structures of Persepolis were built.

In any event, this Persian cleverness with air and wind, a cleverness of unknown antiquity, constitutes our first glimmer of evidence. Our second glimmer is found in Islamic travelers' commentaries about one of the least attractive corners of Persia, Seistan. This is an eastern border district where the Helmand River, which rises in the highlands of Afghanistan still farther to the east, comes to a pathetic end in a vast marsh surrounded by desert. "The land is flat," wrote one early traveler, "and one sees no mountain . . . strong winds prevail, so that, because of [the winds], mills were built rotated by the wind." A later visitor says of Seistan ". . . there the wind turns mills that pump water from wells to irrigate the garden."

Of course, the very fact that I can cite quotations like this puts us back in the realm of written history. But have we reached a point earlier in history than the Yorkshire windmill of A.D. 1185? Very little, alas. The two travelers who described Seistan made their journeys in about A.D. 950. So, at best, this Persian excursion has added only two centuries or so to the historical antiquity of the windmill. Hang on a minute.

Still a third Islamic writer turned his attention to desolate Seistan, although not until the 14th century A.D. What he did, however, was far more than the earlier visitors had. He actually described a Seistan windmill and gave details about its construction. Forget your Dutch preconceptions. The architects of Seistan built mills with narrow upright sails that made a vertical air-powered axle rotate. They were equivalents of the "horizontal" water mill! And how were these "horizontal" windmills housed? Inside wind towers that were modified for this very purpose.

Horizontal windmills could still be found in China not too many years ago. Like the horizontal mills of Seistan, the Chinese ones were used mainly to pump water rather than to grind grain. In the giddy altitudes of Tibet the pious may still construct miniature horizontal windmills in the form of prayer wheels, to accumulate virtue rather than to do useful work. The point here, of course, is that no wider a conceptual gulf separates a horizontal windmill from a vertical one than separated the horizontal water wheel from its vertical cousin. A person who sees one can easily imagine the other, and in both examples the vertical version is the more efficient.

So here, at least for the present, we may sum up by saying that the horizontal windmill is historically documented in the East more than two centuries before historians can certify the presence of a vertical windmill in the West. How was knowledge

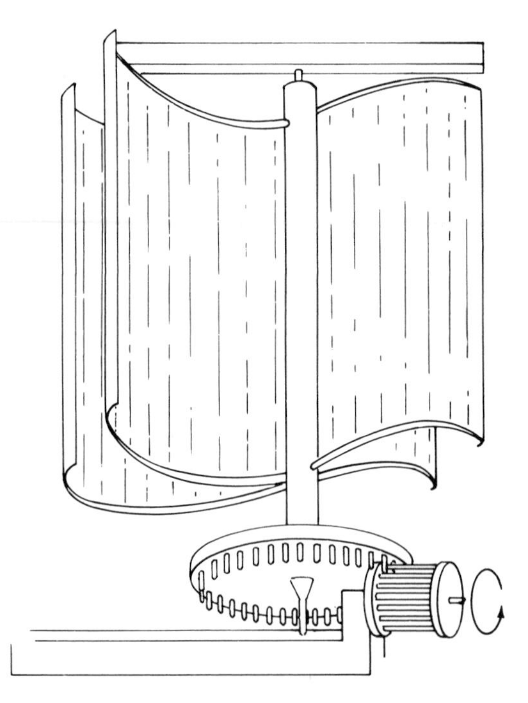

Horizontal windmill, like the horizontal water wheel, was a very early prime mover and may have been the first of all windmills. Unlike the horizontal water wheel, however, it did not lead to anything so dramatic as the water turbine. Mills like this were used in eastern Persia to pump well water to the surface for irrigation; the cog-and-reel gear train shown here could have driven a simple rag-and-chain pump. In China, similar horizontal windmills were also used to raise water for irrigation.

of the windmill principle transmitted from Asia to Europe? Perhaps no knowledge ever traveled west and the vertical windmill was an independent western invention. Or perhaps not. For myself, I think it would be foolish not to admit the possibility that both on the frontier in Seistan and elsewhere among the wind-wise Persians this air-powered innovation, horizontal though it was, had been in use long before Islamic travelers came to take note of it. In any event, I refuse to believe that windmills, vertical or otherwise, were a 12th-century English invention.

Wherever the windmill may have been invented, it was a tremendous success and came into use — primarily as a gristmill but also frequently as a water pump — throughout much of the Old World. We cannot leave this shrewd device for tapping the energy of the restless air without drawing attention to the mechanics of these devices. Like the water wheel, the windmill is essentially a corporation of levers, applying a twisting force to a central axle that rotates in consequence. In the case of the windmill the twisting force is applied through all the levers — the wind sails — simultaneously. (In an overshot water wheel, on the contrary, at any one time about half the levers — the buckets — are dead weight, doing no work.) In terms of the respective fluid and gas that are doing the work, generally water applies more pounds per square inch to the buckets of a water wheel than air applies to the wings of a windmill. This is because, to say it simply if not with precision, water is "heavier" than air. Bear in mind, however, that the superiority of water over air (a calculation that must also take into account the size of the wheel buckets and the wind sails) is far less important than a day-to-day practical consideration. Given a reservoir and a perennial stream, a water mill can grind grain (or turn machinery) every day the year round except in the most severe of winters. On still days or days that are too windy, however, the windmill must stand idle.

Sails, of course, can be used to achieve linear, as well as rotary, motion. The prehistory of ship's sails is difficult to trace;

these thin sheets would soon rot away, even when stored on dry land. Before the weaving of cloth in large sizes made it practical to manufacture canvas (that is, cotton) and linen sails, it is possible to imagine that specially prepared leather sails could have been made for small boats. It is equally possible that the early small-boat mariners, such as the Mesolithic obsidian carriers of Franchthi, would have found paddling (or rowing) weary enough work to make the use of even small leather sails desirable. Such speculation remains, however, wholly in the realm of the imagination. By the time we encounter "prehistoric" depictions of boats — such as, for example, the rock drawings of Scandinavia early in the Christian Era — sailing vessels had been historically documented for thousands of years in the Mediterranean and Asia.

Nevertheless, sailing craft typify man's earliest harnessing of the air, and epic voyages under sail were being logged throughout the world before either water wheel or windmill was a glimmer in Miner's or Miller's eye. The history of sail has been written in detail. Here it needs only the briefest summary and due recognition of its significance in man's pursuit of power over and above his own muscle.

First, then, on the evidence of artists' depictions (the earliest representations that show significant details being Egyptian), the original wind catcher was a square sail, rigged for best running before the wind. The square sail, together with its cousin, the adjustable square sail, and such modified variants as the lateen sail and various triangular jibs and square topsails, dominated the sail plan worldwide for long millennia of aerodynamic ignorance. The phrase "running before the wind" summarizes this ignorance: simple reason led mariners to believe that the wind pushed them along. After all, they know that a gust of side wind makes a ship tip.

The more efficient fore-and-aft sail plan that dominates the world of sail today was not developed as a result of sudden aerodynamic enlightenment. It was a cut-and-try development that won acceptance because of improved performance. Its inventors almost surely thought they were still harnessing the pushing force of the wind, although they were doing so in some novel and not clearly understood manner. It was not until the 18th and 19th centuries that a succession of English windmill engineers gave up the cut-and-try approach to air-power technology. These pioneers began to calculate the action of the wind upon square plates that were fixed at a variety of angles incident to the moving air; the result was that the first dim glimmer of aerodynamic analysis became visible. There is some modest

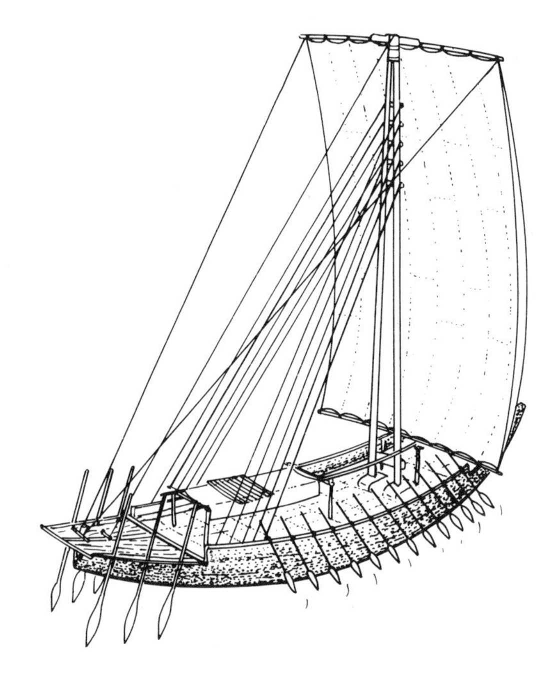

Linear wind power: this restoration of an Egyptian sailing vessel shows that the ship carried its share of levers (26 rowing oars and six steering ones). When the wind was right, however, its great square sail, hung from a bipod mast, must have made the ship travel at a good clip. Since it usually "ran before the wind," it is scarcely surprising that mariners thought the wind was pushing the ship.

irony in the fact that the concept of lift — pioneered by windmill engineers and a commonplace of aircraft design since the 1880s — was alien to mariners until the 20th century.

Yes, that's what a sail does. Just like a wing, it interferes with the normal flow of air to create a low-pressure area on one side. Once such an area is created, the sail moves in the low-pressure direction, in the same way that a wing tends to "climb" in that direction. It is not the wind that bellies out the sail by pushing from behind; the sail bellies out in response to low air pressure in front. What propels the sailing vessel is the reaction to this pressure differential, transmitted by the sail — think of it as an aerodynamic plate — to mast and rigging. Again, to say it simply if not with complete precision, the wind isn't pushing, it's pulling.

When it came to serious marine endeavors, such as naval warfare, the ancient mariners did not put their trust in sail any more than the first steam dreadnought admirals did. Lacking the dreadnought's boilers and pistons, the ancients depended on muscle power. We have all heard of galley "slaves" although, in fact, most rowers in history have been seasonal volunteers who were well paid for their labors. From the time of the first Greek seafarers up until the 17th century, whereas sails were used on long hauls, when it came to combat, all the wind apparatus — mast, spars, and sail — was tumbled down for deck storage. When it was time to strike the enemy the last sprint into battle was under oar.

So here we are back to muscle power again. This is actually apt because, until the heat engine came into being, all those admirable uses of rotary motion we have examined were completely dependent on the muscle power of men and animals where water power or wind power was not available. For example, a word that is quite as indicative of forced labor as "galley slave" is "treadmill." This ingenious, if not particularly attractive prime mover worked by muscle power, both human and animal. Remember the water lifter at Rio Tinto? Who made the wheels turn? Men did, treadmill fashion. In A.D. 1346 — less than 900

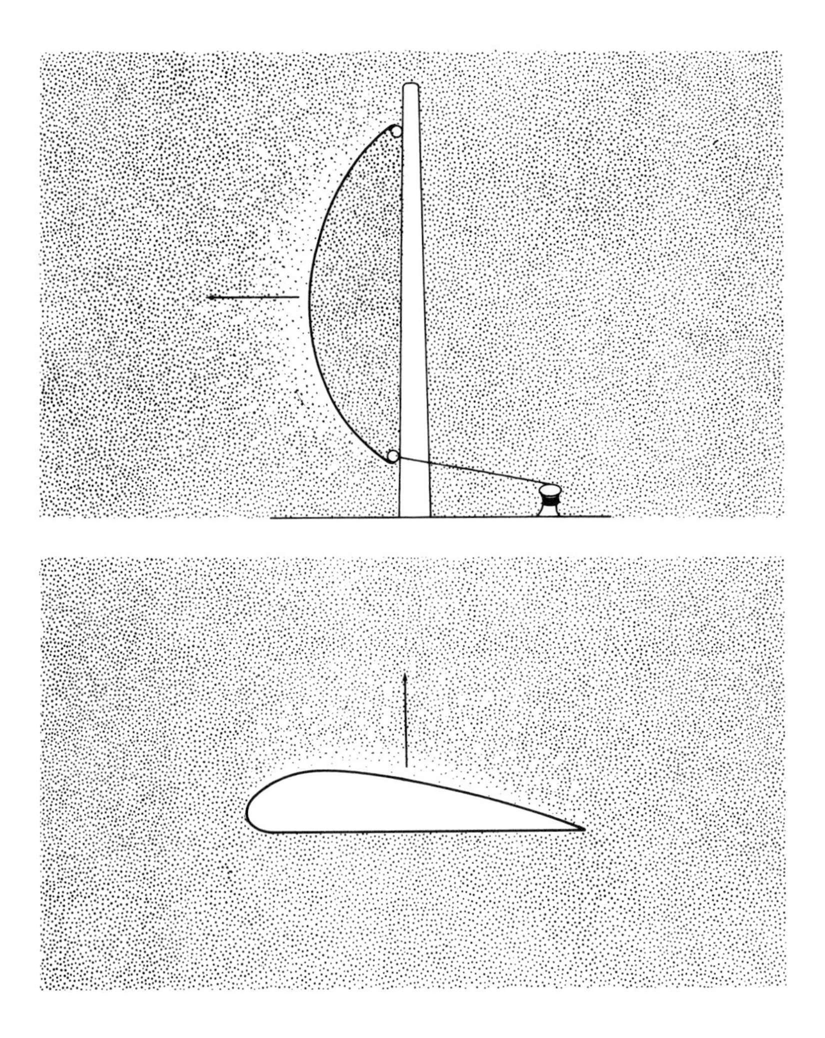

A question of "lift": the force acting on a sail, which makes a ship move forward, is not the push of the wind from behind but the reduced air pressure over the front of the sail area (*arrow*). Exactly the same is true of the wing of an airplane: its curved upper surface forces the air that flows past it to travel faster than the air flowing past the flat bottom surface. This reduces the air pressure on the top of the wing, so that the wing has "lift" (*arrow*) and rises.

years after the fall of Rome and almost 150 years before Columbus's voyage to the New World — a great crane was built in the German town of Lüneburg, a port on the river Ilmenau 40-odd kilometers inland from Hamburg, to handle waterfront cargo. And what hoisted the bales at Lüneburg? Human muscle, applied to a pair of treadwheels.

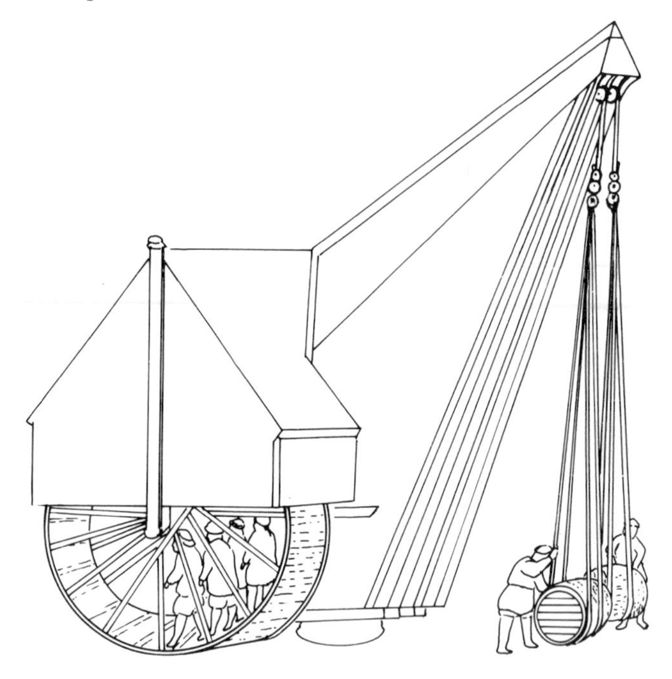

Well, back to the old treadmill. This crane, like the one at Lüneburg, operated by manpower. This is the medieval crane at Bruges (now in Belgium, then part of the Holy Roman Empire). The contemporary artist, whose painting was the basis for this line drawing, did not have much feeling for machinery. He has the four men so positioned that they would have to walk backwards to turn the wheel.

The Muse of History may have developed a taste for irony. Having seen centuries of service, the Lüneburg crane was replaced in 1726 by a close duplicate, also man-powered. The crane operators maintained careful records and so we know the maximum load the 1726 crane ever lifted. The event took place on the 13th of August, 1840, and the load was a British-built steam locomotive: the first external-combustion prime mover to be delivered to the newly built Brunswick railway line. Landed by manpower.

So men were working treadmills well into the age of steam. This is not really so surprising. Before little motors came along (which was sometime after big motors had put most water wheels and windmills out of business), human and animal muscle continued to do most of the prime moving. Consider that triumph of cottage industry, I. M. Singer's sewing machine. To this day, in rural areas beyond the reach of electricity (or in areas that are simply poor) you will find Mr. Singer's machine still powered by human muscle — specifically, leg muscle — as it was when he invented it in the mid-19th century.

Manpower was preferred over animal power in treadwheel devices such as cranes, where fine performance was required: not only for cargo handling but also for construction work. To return to the past from this brief 19th-century excursion — specifically, to the early Iron phase of the Age of Metals — consider how the Greeks built the Parthenon in the later half of the 5th century B.C. Those fine fluted columns that support the roof are not monoliths like Cleopatra's Needle. They are composed of individual sections ("drums"), usually 11 to a column, each section weighing eight tons or so. Perhaps the bottom section of each column was wrestled into place manually, although this seems doubtful. The other sections were surely lifted into place by means of portable cranes. To set each section accurately on the one below — the top section needed to be raised some 30 feet into the air before being positioned — is not something one does by saying "gee" or "haw" to a team of mules. With tolerances as close as these one wants men at work, preferably on treadwheels. Please note: no one can prove that the builders of the Parthenon and the other drum-column buildings in ancient Greece used cranes. But what was the alternative? Historians were made happy when some 700 years later a treadwheel crane was actually depicted in a 2nd-century A.D. Roman tomb relief.

For donkey's work, of course, there are always donkeys. You have already seen the evidence of man's progress as a domesticator of wild animals, beginning with the appearance of the first tamed dogs of the Old World and the New and with the change in the sheep at Shanidar. The majority of the prime-mover animals — ox, buffalo (and relatives), horse, donkey, and

Animal workers: the dog's role as a servant of man, as has been seen, began many thousands of years ago and continues to this day. In a great 16th-century work, *Concerning Metals,* Georg Bauer [Agricola] noted that miners of the period used dogs as pack animals for carrying ore. The scene here is from a woodblock in his book.

mule — came into service somewhat later and, like the dog, have not escaped service yet. I need not tell you what you already know: that these large animals, along with the elephant, the two varieties of Old World camel, and at least one New World camel, the llama, can be found in use as work animals to this day and that most of them are not yet freed from labor beneath the yoke or between shafts. Neither transportation nor agriculture is as yet wholly mechanized. Nonetheless, animal traction and portage are on the wane.

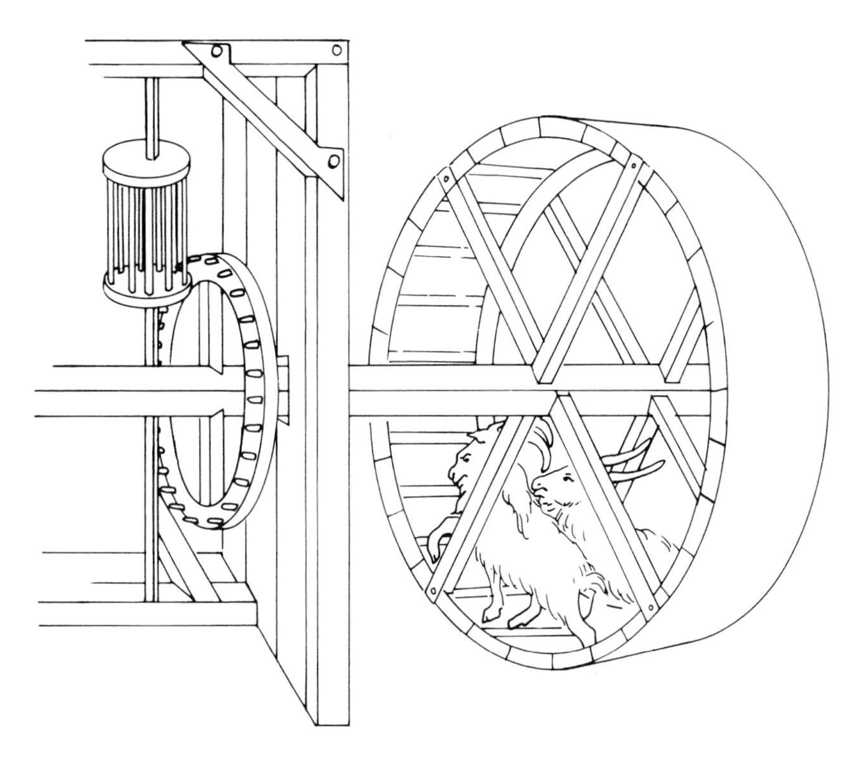

Two-goatpower mill: how the rams were enticed to keep walking uphill is not clear. If they kept it up, however, the toothed gear inside the shed would make the cage gear rotate and its rotation would drive a millstone (*not shown*) in the second story of the shed.

Instead, I want to direct your attention to the role of these domestic animals (and I now add the dog to the roster) as operators of machinery. This role, barring continued employment of a few mill donkeys and oxen, is no longer known today although it ended only yesterday. Which were the first of these machinery operators? Probably the same species that are now the last: the donkeys and oxen that tread a monotonous circle, turning the grindstones of small mills that produce anything from cider to many kinds of flour to soybean curd. *And* (emphasis intentional) working pumps. Not only water lifting for irrigation but specifically water pumping, as with a rag-and-chain pump, to keep mine workings as dry as possible. I will return to this topic in the next chapter. But the main prime mover in the western world was the horse, and the machinery he powered was steadily improved in design until, by the middle of the 19th century, a single cast-iron "horse engine" could be used to thresh grain, to chop fodder, or to pump water, according to the owner's wishes.

In England alone, by 1870, at least 21 manufacturers pro-

Four-horsepower prime mover was built in the 19th century by an Italian who lacked faith in steam engines. Records do not show that this super-treadmill attained any particular popularity.

duced horse engines. Their power trains, involving as many as four gear-ratio changes, transformed the horse's ambling three revolutions per minute around the central pivot into shaft speeds approaching 300 r.p.m. In the U.S. and Canada, meanwhile, where the prairie wheatfields yielded staggering quantities of grain for threshing, horse-engine threshing machines powered by as many as 24 animal prime movers were offered for sale to the frontier farmers in the 1880s.

In Biblical times, we remember, it was the oxen that treaded out the corn. As the 20th century approached in the western world threshing had become a matter of horsepower and gear ratio. The transition to the energetics of today was very nearly at hand. The first installation of an electric light in a lighthouse had been made even earlier, in 1862, at Dungeness Head, overlooking the English Channel near Dover.

Its dynamo was spun by a steam engine, a lineal descendant of the first practical heat engine, the external-combustion device pioneered by Thomas Newcomen in the first decade of

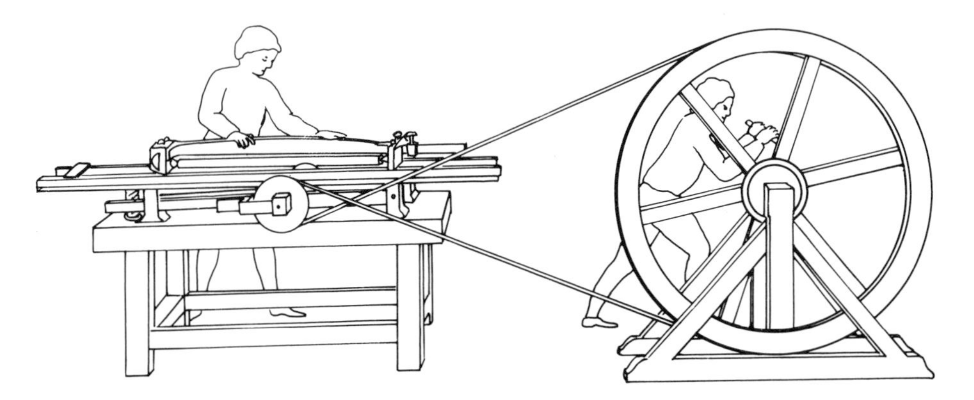

One-manpower crank: this device, with its enormous flywheel, made a little wheel on the workbench spin like the dickens. Its axle, in turn, spun a grindstone that put a smooth finish on the musket barrel that the second worker is pushing back and forth. This 18th-century machine shop is illustrated in Diderot's *Encyclopédie*.

the 1700s. Meanwhile, a far piece across the Channel and inland, one Nicolaus Otto was well on his way to perfecting the second practical heat engine, an internal-combustion device (his one-cylinder, four-cycle model reached the demonstration stage one century after the American Declaration of Independence). Mankind was destined to depend on animal power for decades to come, as many men still do, but new means of harnessing energy were in the wings and already crowding onstage. The era of mechanical rather than muscular power had begun with the building of the first heat engine. Let us see how this happened.

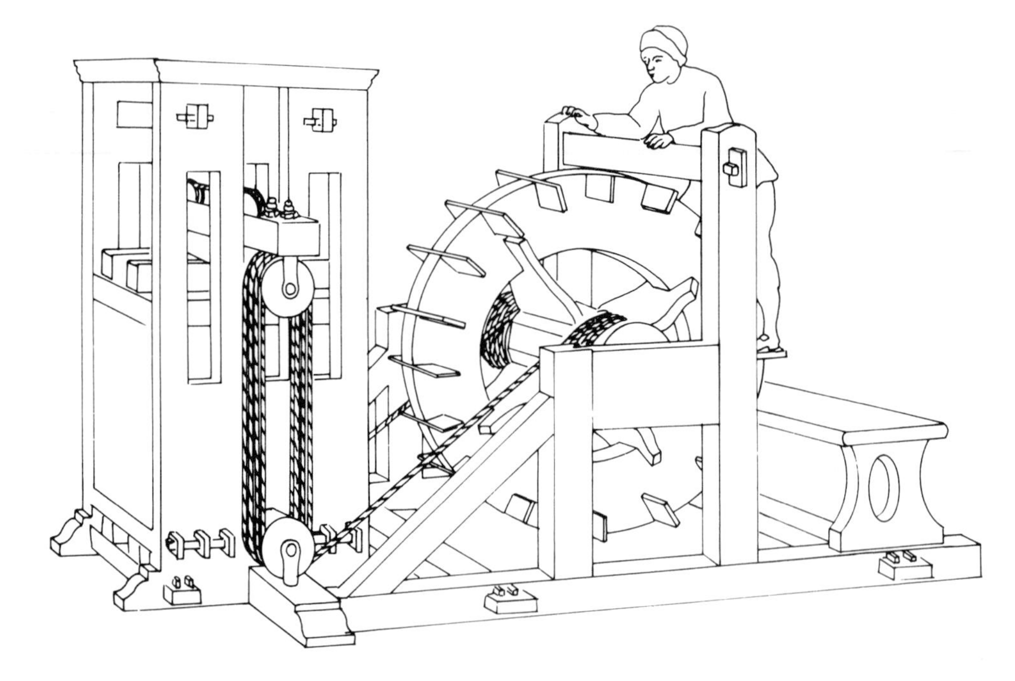

Two Greek "simple machines" combined, utilizing one manpower as the prime mover. The wheel-axle treadmill slowly but surely reels in two ropes that run through paired pulleys. As the multiplied force of the pulleys makes the flat bed of the press descend, the contents of the press come under enormous pressure. This is another device illustrated in the *Encyclopédie:* it was used for bailing paper.

CHAPTER

10

The Rise of the Heat Engine

It all began with man's thirst for metals. If a miner starts to work in a valley and drives his ore-seeking shaft into the hillside, water gives him little or nothing to worry about. To be sure, his digging may tap what are called aquifers, or water-bearing levels, below ground, just as a well-digger hopes to do when he sinks a shaft straight down from the surface. If, however, his shaft ascends a little as it advances, his mine will drain itself; whatever water he releases will simply run down the slanting shaft and pour into the valley.

But what if friend Miner is pursuing a vein of ore that trends downward? Or if he starts at ground level in the first place and sinks a vertical shaft like a well-digger? Either way, when he hits a water-bearing zone he has to quit or start bailing. And bailing by the bucketful will not help much. As we have seen, in the copper mines of Rio Tinto at least as long ago as when the Romans were the rulers of Spain (and perhaps for centuries before that under the Phoenician colonists of North Africa), bailing was done by a succession of man-powered water wheels or, more properly, water lifters. Very useful those 12-foot wheels were, too. But expensive! The excavation of each wheel chamber was a major enterprise. As some poor sweating rock-chopper doubtless remarked, "We could use a pump."

Wily old Miner and his thousands of cousins — for in Western Europe alone during the era of imperial Rome the metal-ore mines numbered in the hundreds — were of the same mind.

Alas for his sake and the sake of all his kin, a force pump that might have solved their problems had already come and gone and been forgotten. It had been designed by a Greek artificer of the 3rd century B.C., one Ctesibius by name, an honored member of the colony of Macedonian overlords that had been planted upon Egypt by Alexander the Great. (Our old acquaintance and analyst of the five simple machines, Hero of Alexandria, was one of Ctesibius's pupils.) Because of the loss of Ctesibius's force pump, the miners of Greece and Rome and their successors for centuries had nothing better to use than suction pumps.

What distinguishes a suction pump from a force pump? With a force pump, given strong enough construction and enough power you can push water straight up just as high as you please. With a suction pump, even a 100-percent efficient one, you can lift water no higher than 33 feet. Why? This observed limitation puzzled even Galileo.

To answer the "why," we must, for a start, recognize that the word "suction" is a downright lie. No such thing as a sucking force exists; there are only different ways of creating higher and higher (or emptier and emptier) vacuums. Once a vacuum is created, it is the ocean of air that surrounds each of us that does the subsequent work. And it does so by pushing. In the case of the misnamed suction pump, once you have created a partial vacuum in a pipe (by pulling up a tight-fitting plunger placed inside the pipe) you are able to use the atmosphere as a working force. Specifically, you have at your disposal the pushing force of the weight of a column of air of the same diameter as your pipe, stretching from the below-ground surface of the mine water you are trying to pump all the way up to the stratosphere above your head. The column is even taller than that, but by then the air has grown so thin that it no longer adds much to its total weight. For ease of mental arithmetic, let's call the total distance 100 miles, or 528,000 feet above mine water. The weight of the first 1,000 feet of this invisible column is far greater than the weight of any one of the 527 1,000-foot units of atmosphere stacked on top of it. Nevertheless, the cumulative effect of all that air is to press down on the first fractional millionths of an inch resting on the mine water with a total load of 14.7 pounds per square inch. The same load is pressing down, right now, on thee, me, and the nearest tree.

For simplicity, consider that your mine-pumping pipe has a diameter that accomodates exactly one square inch of water surface. When you pull up the plunger (thereby creating a vacuum in the pipe) the mine water will rise up your pipe until the weight of the water column (in a 100-percent efficient system) comes to 14.7 pounds. Now, how tall is a square-inch column of water that weighs 14.7 pounds? You've got it! Thirty-three feet!

Try it with mercury, a fluid much heavier than water. How far will a column of mercury be pushed up an evacuated tube by a column of air of the same diameter? Less than three feet, or 30-odd inches. The exact distance depends on the atmospheric pressure, that is, on the ever-changing weight of the air column. This is why the variations in atmospheric pressure, such as weather forecasters discuss, are usually measured in inches of mercury. An atmospheric (the forecaster will say "barometric") pressure of 29 inches or less is deemed very low, and a 31-inch pressure or above is very high.

What if we chop off the heavier part of the towering air column? Let's throw away the first 50,000-odd feet of it by ascending in a balloon to that altitude (being sure to carry along plenty of oxygen and to wear heated clothing). With the air column reduced to a length of only 90 miles what happens? The mercury in your barometer now rises only three inches high and the water level in the model mine pump you have brought along will climb no higher than 3.3 feet, no matter how hard you pump. This is because the 90-mile air column above your head weighs only 10 percent as much as the 10-mile air column under your feet. With these facts about the power of atmospheric pressure tucked under your belt, let's go back to the mines.

In a deep mine, if you are willing to construct a series of tanks, you can keep the bottom dry with "suction" pumps. To use round figures, if your shaft is less than 280 feet deep you will need only 10 such "suction" pumps and nine tanks in the shaft

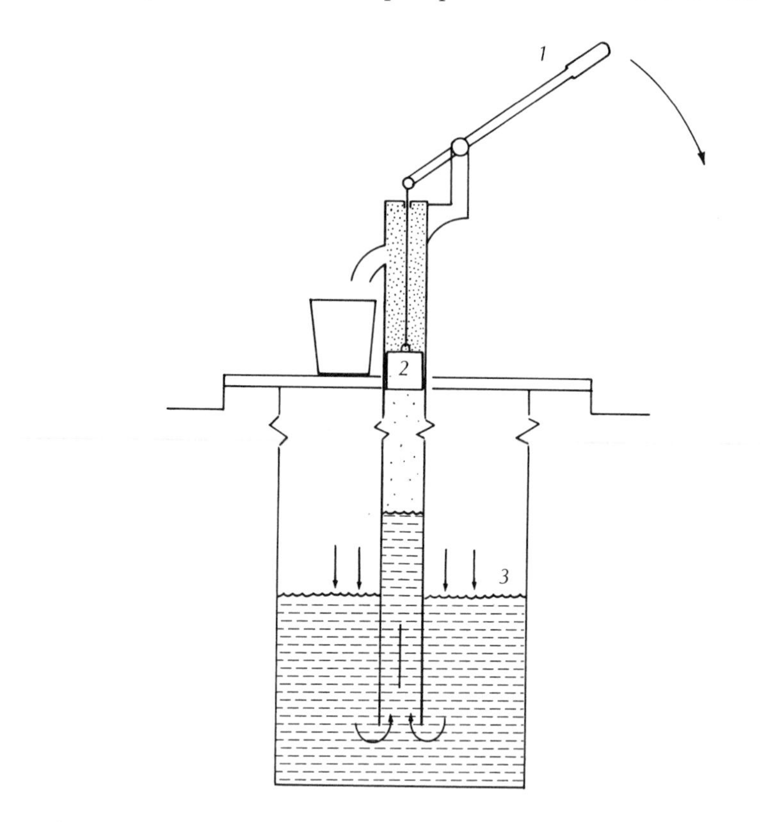

The suction pump is an "atmospheric engine." A push down on the handle (1) draws a piston (2) up the pump tube, forming a partial vacuum in the tube below the piston. The pressure of the air on the surface of the well water (3) is 14.7 pounds per square inch; this pressure forces the well water up into the evacuated pump tube. How far? Far enough to have each of the square-inch-size columns of water in the pump tube weigh 14.7 pounds. How tall *is* a square-inch-size column of water that weighs 14.7 pounds? It's 33 feet tall. That is why no suction pump can raise water more than 33 feet high. Notice that all the pumper does is to create the partial vacuum in the pump tube. It is the weight of the air that pushes the water up.

— and teams of 10 strong men pumping in shifts day and night! Time was, of course, when animals, slaves, or prisoners were available for this kind of tiresome endeavor and, as we have seen, that time was not so long ago. Nevertheless, rather more than 250 years ago another way to pump — utilizing the exact same weight of air overhead that gives the "suction" pump its power — came into existence.

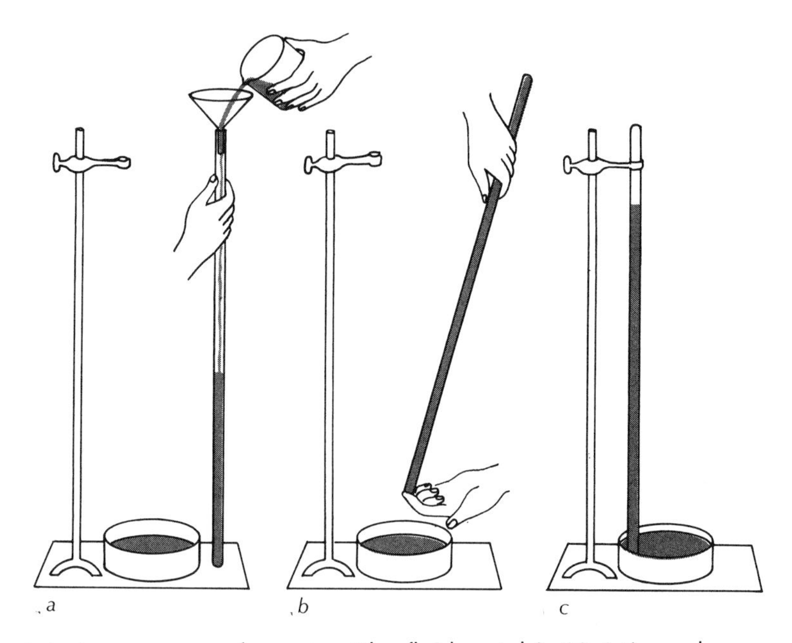

Substitute mercury for water. The fluid metal is 13.6 times denser than water and so a square-inch-size column of mercury weighing 14.7 pounds in only 30 inches tall. Set such a column up in a 36-inch-long tube of glass: Voilà! A barometer. Half-fill a shallow dish with mercury; also fill the tube with mercury (*a*). Sealing the open end of the tube with your finger, invert the tube and put that end in the mercury pool in the dish (*b*); remove finger. Six inches of mercury will run out of the tube into the dish, leaving a vacuum at the sealed end of the tube. Secure the sealed end in the support clamp unless you want to stand there day and night. Hereafter, changes in atmospheric pressure will make the column of mercury rise or fall.

Known as an "atmospheric" engine, it used that form of hot, expansive water vapor we have all called "steam" for centuries now to create a sudden near-vacuum inside a very large cylinder. Once the vacuum was formed, atmospheric pressure forcefully pushed down a piston that was poised at the top of the cylinder. This power stroke downward was converted (by means of a "walking beam") into a pull upward and this pull, or upstroke, created vacuums in a whole series of underground "suction" pumps.

Because this vast combination of cylinder, piston, walking beam, and pump rod took its supply of steam from a boiler that rested over a fire some distance away, it must be considered the first practical heat engine, even though the unheated atmosphere actually did the work. And because the source of the steam was outside the cylinder, it was also an external-combustion engine. (An engine that is driven by the sudden expansion of a fuel, such as gasoline vapor, set on fire *inside* a cylinder is a heat engine of the internal-combustion kind.)

It seems doubtful, however, that Thomas Newcomen, whose first attested atmospheric engine began pumping a coal mine dry in Staffordshire, England, in 1712, gave much thought

Newcomen's 1712 atmospheric engine, which powered a series of suction pumps, was a kind of giant suction pump itself. A big cylinder, fitted with a piston (1, *a*), was connected to a boiler (*b*) and a reservoir of cold water (*c*). Steam was allowed to fill the cylinder (2), after which a spray of cold water from the reservoir (3, *a*) condensed the steam, producing a near vacuum. Atmospheric pressure then forced the piston to the bottom of the cylinder (4), pulling down one end of the walking beam and pulling up the other (4, *a*). The upward motion pulled up the pistons in a cascade of suction pumps arrayed in the mine beneath the engine, removing water from the mine (4, *b*). A counterweight (4, *c*) repositioned the rocking arm, raising the engine's piston to the top of the cylinder, and the sequence was repeated. Newcomen's first atmospheric engine pumped 120 gallons of water per minute out of the Coneygree coal mine.

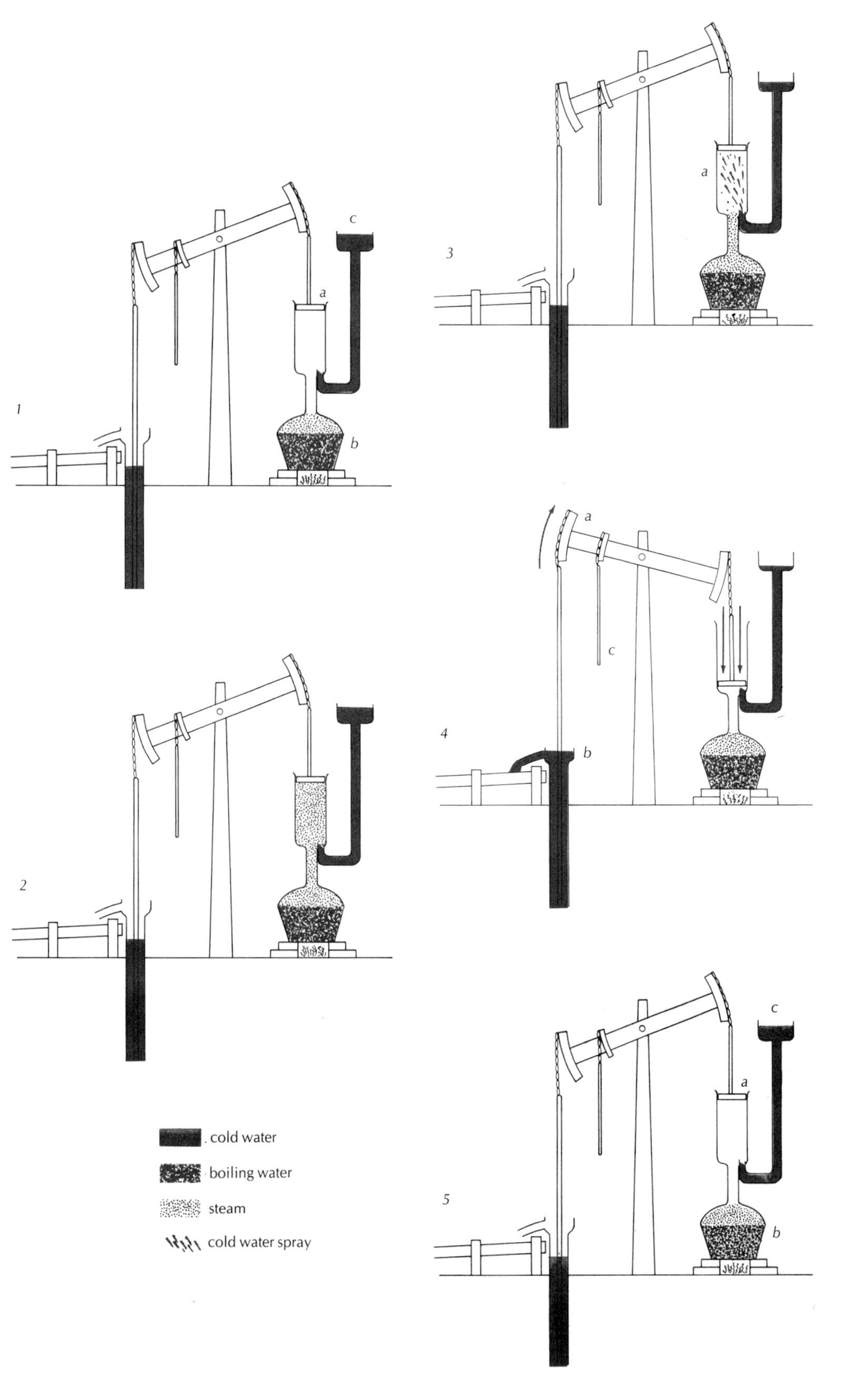
1
a
b
c
2
3
a
4
a
b
c
5
a
b
c
cold water
boiling water
steam
cold water spray

to the combustion aspect of his engine. After all, the steam was merely used to create a vacuum; the atmosphere did all the work.

Newcomen's 1712 engine at the Coneygree Coal Works pumped 120 gallons of water per minute out of the mine, raising it by stages from a depth of 150 feet, and chugged away at a rate of 12 strokes per minute and 10 gallons per stroke. For decades and even centuries thereafter other mines round the world would be kept dry by means of animal and human muscle, but never again, from this point early in the 18th century onward, was such use (or misuse) of muscle power absolutely necessary. With the construction of Newcomen's engine, man had reached a completely new level of achievement in the mastery of energy.

Hitherto, save for muscle, only the natural forces of wind and water had been at man's disposal as a source of work. What about fire? To be sure, fire had been his for at least half a million years, but its contribution to mankind had been confined to increasing human efficiency indirectly: lengthening the day, improving the microclimate, predigesting food, turning clay into pottery and ores into artifacts. Now, by providing steam for Newcomen's engine, fire had become directly involved in doing work. Read on.

Since there are 60 seconds to the minute, the Coneygree engine was raising two gallons of water per second from a depth of 150 feet. Two gallons of water weigh 16 pounds. Expressed as horsepower (550 foot-pounds per second) this performance approaches four-and-a-half horse. Allowing for friction and other inefficiencies, the blaze under Newcomen's boiler must have been releasing energy in the form of heat at a minimum rate of 2,000 calories (1,000 calories equals 3,082 foot-pounds) per second. I suspect that, actually, the boiler fire was many times more energetic than this. The heat loss involved in operating an engine like this, or indeed any kind of heat engine, is staggering.

Efficiency, however, was of scarcely any importance here. Far more coal came out of the mine than was burned under the boiler to keep the mine dry. Meanwhile, for the first time in human history, the chemical energy stored in fuel was being combined by combustion with the chemical energy of atmospheric oxygen and transformed into radiant energy — heat — that activated a mechanism capable of delivering more work than a four-horse team! Milestone! Enormous milestone, enormously far along the path we began to trace over two million years ago.

Just before we leave the coal mines of England and pass from atmospheric engine to steam engine and the various energetic applications that produced a host of ipsimobiles, or self-propelled vehicles, let me offer you a few trivial facts and figures about English coal and coal production, in Q. & A. fashion:

Why did (and do) the English call coal "sea coal"? Because, when organized mining of coal began in the 13th century A.D., the mine shafts extended under the estuary of the Tyne River (near Newcastle) and so the coal was, so to speak, mined at sea. In those days, when one simply said "coal," one was speaking of a forest product: charcoal, which, next to firewood, was then the most common English heating material. To be sure that one would receive the curious black rocks, rather than a sack of charcoal, one specified "sea coal," *carbo maris*. Today there are better-than-ever reasons in England for using the term sea coal. Of the 150 million tons of coal now mined annually, better than half comes from mine shafts that run for thousands of yards out under the North Sea.

What about the derogatory phrase, "carrying coals to Newcastle"? By the first quarter of the 14th century (in A.D. 1325, to be specific) enough Tyneside coal was being mined to allow the surplus to be shipped to France aboard vessels sailing from Newcastle. Clearly, only a booby would try to transport a commodity to a place that already exports it.

What did steam pumping do for the mines of England? Statistics for the 17th and much of the 18th century are indifferent. This was a period of mine expansion, however, and not only Newcomen's pump but others would appear, most notably after 1782 those of James Watt. Yet by 1770, more than a decade before Watt's engine, annual coal production in England had passed the level of one million tons. (By 1900, the first year of our own century, the annual rate was 46 million tons.)

The folklore of technology often emphasizes two themes. One of these is the commonplace observation that leads a superior mind on to some key innovation; the other tells of the dedicated innovator whose life ends miserably while the world goes on to enjoy the fruits of his invention. James Watt, born in Scotland in the winter of 1736, provides an agreeable exception to both themes. Although folklore attempts to make young Watt realize the potential of steam while watching a kettle boil on the home stove, not the slightest reason exists to believe that such a flash of genius ever took place. As you will see, Watt learned all he needed to know about steam power in a quite straightforward way. As far as coming to a bad end is concerned, Watt enjoyed a full, prosperous, and honored old age and died peacefully at the age of 83, thereby defying the folklorists a second time in one life. During that one life, moreover, he saw the kind of steam engine that he first perfected develop into an entirely novel prime mover both at sea and on land.

Apprenticeships were the rule in young Watt's day, the hard road whereby a youth without private means or family connection might educate himself and learn a bourgeois skill. In Watt's case this was arranged in London, a city far from his birthplace in Greenock; at age 19 he was apprenticed to one John Morgan, a maker of scientific instruments. A year in gritty London very nearly killed the young man and Watt returned to Scotland for the sake of his health without completing his apprenticeship. In Glasgow, where he sought work, his record as a failed apprentice was a black mark and none of the Glasgow guilds would admit him.

Fortunately for the young man, who by this time had turned 21, Glasgow College was not subject to guild discipline and so, in 1757, Watt was appointed instrument maker to the college. A task that fell to him in this capacity some seven years later was to repair a working model of a Newcomen atmospheric engine in the college collection. Analyzing the model, Watt recognized the inefficiency involved in first cooling and then reheating the engine's single cylinder. For each powerful cold downstroke, entirely too much steam was wasted during the subsequent upstroke.

Rather than tinkering straightaway, Watt began his work at the theoretical level by measuring the density and pressure of steam over a range of different temperatures. He concluded that, in a theoretically efficient engine, the steam entering the cylinder during the upstroke should be at as low a temperature as possible. In this way the condensation of the steam (the action that produced the vacuum within Newcomen's cylinder and allowed atmospheric pressure to power the downstroke) would waste a minimum of heat. Watt's corollary conclusion was that the cylinder should not undergo alternate heating and cooling but should always remain at least as hot as the temperature of the entering steam during the upstroke. Theoretically, all this was fine enough, but how could these goals be met in a practical world?

If Watt's answer was not a stroke of genius, it was at least ingenious: don't condense the steam inside the power cylinder at all. Instead, use a separate condenser, which would always be kept cold, just as the cylinder would always be kept hot!

Here is how Watt's system was eventually to work. We will start with the upstroke of the piston inside the cylinder. Whoosh! In comes the hot steam, billowing into the space under the rising piston. Off to one side now, an air pump is sucking away, creating a partial vacuum in a large, airtight cold chamber. Whump! The piston reaches the top of the cylinder and a valve closes off the steam line. Instantly another valve opens — this one in the line running between the hot cylinder and the cold chamber. Whish! The steam rushes out from beneath the piston to replace the air that has been pumped out of the cold chamber; the low temperature of the chamber turns the steam into water.

With the steam beneath it vanished away, the piston no longer has a counterbalance against the weight of the atmospheric column above; it starts its descent: the power-producing downstroke. Whump once more: the piston has reached the bottom of the cylinder. The valve to the cold condensing chamber closes and the valve connecting the cylinder and the steam boiler opens; whoosh! A fresh charge of steam begins to pursue the rising piston again. Meanwhile, the condenser side of the system has stayed cold and the cylinder side hot, vastly increasing the efficiency of the engine.

Secret of Watt's improvement over the Newcomen atmospheric engine was not to condense the steam in the cylinder with a cold-water spray but instead to get rid of it in a separate space — a condenser. The condenser was kept cold by a water bath whereas the cylinder stayed hot all the time. Sequence begins (1) with the piston (*a*) at the top of the cylinder (*b*) that has been filled with steam from the boiler (*c*). The water bath (*d*) houses the condenser (*e*) and a dual-purpose air pump (*f*) with a flapped piston. When (2) the boiler valve (*a*) is closed and the two valves at the water bath (*b, between cylinder and condenser and c, between condenser and air pump*) are opened, the lowered air pressure in the condenser permits the steam to escape from the cylinder. The power stroke is about to begin. Halfway through the power stroke (3), steam in the cold condenser has turned to a puddle of water. Flap open, the air-pump piston descends without resistance. At the bottom of the stroke (4) the water in the condenser drains into the air-pump cylinder, covering the top of the air-pump piston (flap closed). Meanwhile, at left, the pump rod of the mine-shaft pump has reached the top of its stroke and water accumulated in the mine, which this entire operation is meant to keep dry, is gushing out of the pump spout. The counterpoise next to the pump rod (*a*) now goes to work; from here on until the next power stroke its lever action will help raise the two pistons on the right side of the fulcrum to the top of their respective cylinders. At (5) the process is halfway completed. Valves *a* and *c* are open, valve *b* is closed; steam is filling the cylinder and condenser air pressure drops. At the top of this recovery stroke the cycle will be repeated. Note that Watt's own ingenious system of automatically opening and closing valves, with striker bars on the air-pump counterpoise, has not been shown in this series of simplified diagrams.

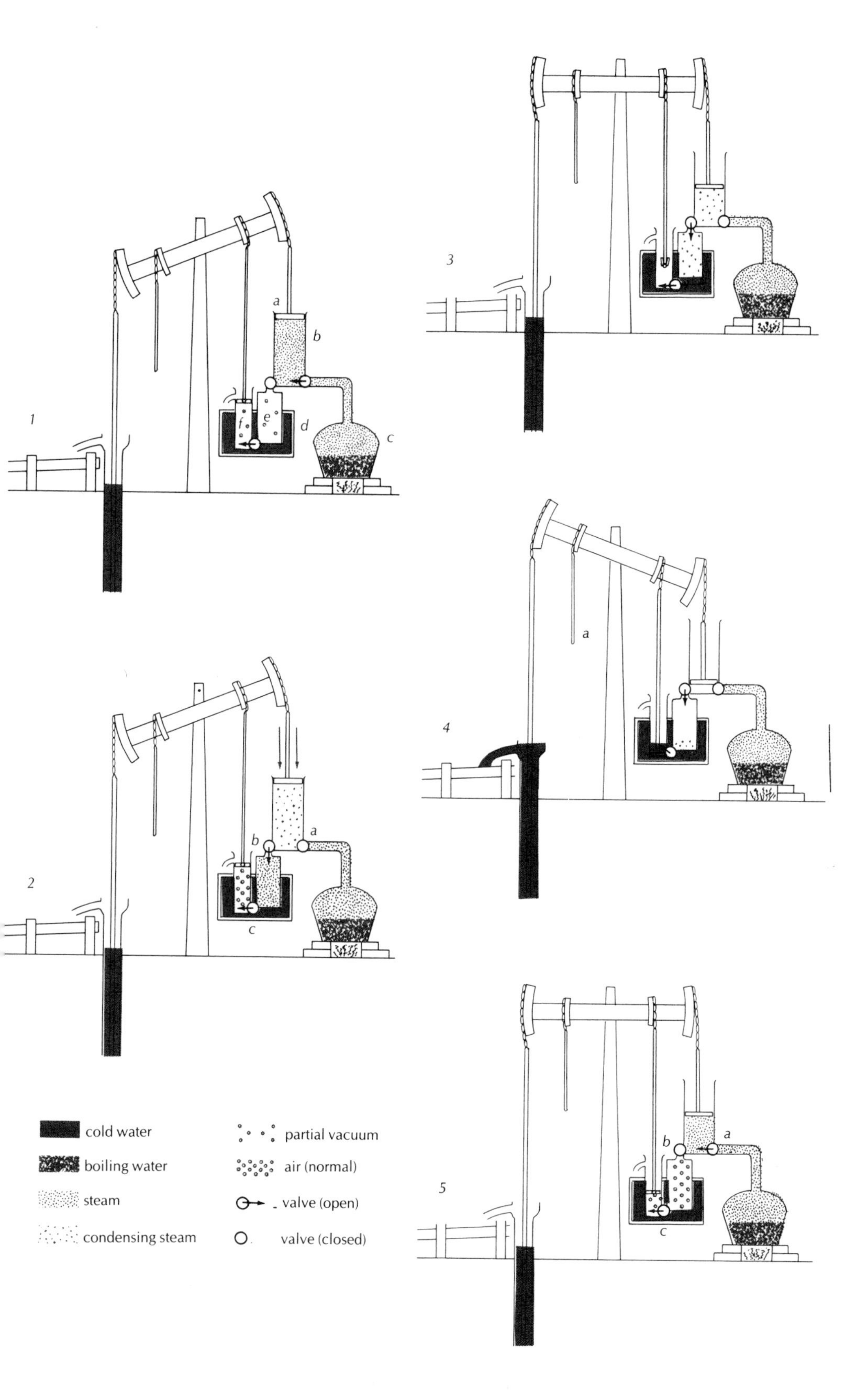
1
a
b
c
d
e
f
2
a
b
c
3
4
a
5
a
b
c
cold water
boiling water
steam
condensing steam
partial vacuum
air (normal)
valve (open)
valve (closed)

Starting with this new concept, Watt tinkered up his own improved model of the old Newcomen atmospheric engine and demonstrated its efficiency. He applied for a patent in 1765; it was granted four years later. The owner of a local ironworks, one John Roebuck, offered to finance the construction of a full-scale Watt engine in exchange for a two-thirds share in future proceeds. Watt accepted the offer and undertook to make his own living as a surveyer and civil engineer during the time it would take to construct the machine for testing at Kinneil, Roebuck's home.

Before much progress could be made, John Roebuck's business failed. Fortunately for Watt, one of Roebuck's friends was a very successful Birmingham engineer and manufacturer, Matthew Boulton, who was quick to see the potential of Watt's engine not only in general but specifically for use in his own factories, which then depended on water-power prime movers. Boulton was also one of Roebuck's creditors and so, in 1773, he took over Roebuck's two-thirds' interest in the Watt engine and in 1774 Watt moved to Birmingham to begin long years of what today we would call research and development. Watt's patent was due to expire in 1782, but one year after he joined Boulton a 25-year extension of the patent was granted and at last the indominable Scot had a little breathing time. Unlike Roebuck, who had pressed Watt for quick results, Boulton — once the patent was extended — encouraged the inventor to work at his own pace. Of the first contract undertaken by the new firm of Boulton & Watt (a 50-inch-cylinder pumping engine for a Cornwall coal mine), Boulton instructed his partner, "Don't let the engine make a single stroke until certain that it will work without a hitch, and then, in the name of God, fall to and do your best."

The big pumping engine was a success: In terms of economic efficiency, it did the same work as a Newcomen engine with the consumption of only one quarter as much coal. By 1783, less than a decade after Watt's departure for Birmingham, all but one of Newcomen's engines in Cornwall had been replaced by Boulton & Watt steam-condenser engines and the firm's monopoly still had 17 more years to run.

Watt invested his first nine years in Birmingham not only in building engines but also in devising general improvements. In 1781, because someone else already held a patent on the simple crank-and-flywheel method of converting reciprocal, walking-beam motion into rotary motion, Watt patented five bizarre alternative conversion systems; one of these — the planetary gear system — was actually used on his first big engines. In 1782 he applied for a patent covering the principle of a double-acting cylinder and the concept of admitting only a brief rush of high-pressure steam into the cylinder and thereafter letting the natural expansion of the decompressing steam do most of the work. A patent in 1784 added a simple connecting-rod guide to the engine. By then Watt had also devised two key refinements to the engine: a throttle in lieu of an "off" or "on" control, and a centrifugal governor that automatically closed the throttle when the operating speed of the engine exceeded a set level. (This latter gadget, by the way, was the first application of what today we call "feedback" control to a heat engine.)

Still another Watt patent, also applied for in 1784, described the use of his steam engine as the driving force of a locomotive vehicle. Watt did not pursue this notion any further, however, and 16 years thereafter, when his original 1769 condenser patent at last expired, Watt retired from business, age 64, giving his one-third interest in Boulton & Watt to his two sons. He died 19 years later, in 1819, secure in the life of an honored country squire even if, perhaps, a little disappointed by being unable to perfect a mechanical means of duplicating sculptures. By the time of his death steamships were regularly crossing the English Channel and one sailing vessel equipped with an auxiliary steam engine to drive paddle wheels — the *Savannah* — had even crossed the Atlantic using both steam and sail. Richard Trevithick had begun selling steam-train rides to London sensation-seekers more than 10 years earlier, whisking them round a circular track at 12 miles per hour. By the time of Watt's death, such other steam locomotives as William Hedley's *Puffing Billy* and George Stephensen's Killingworth machine were hauling freight in the English countryside. Not least, the old gentleman's name was destined to attain undying currency as the term for a unit of power, the joule-second or watt, most commonly used today as a measure of an entirely new source of power: electric energy. The next time you look to see how much electricity you consumed during the month, reckoned in kilowatt-hours, spare a thought for the successful instrument maker of Glasgow.

So much for external combustion: the system that raises the temperature of a "working fluid" by applying heat at some distance from the cylinder-and-piston "engine" where the working fluid actually performs its work. The external-combustion system is still in use today, not only in the few remaining steam locomotives scattered round the world but also for the generation of electric power. In fact, today's newfangled nuclear power plants are nothing but fancy steam generators of the same external-combustion kind, using the steam to spin a turbine. The difference is only that the external heat source is uranium fission rather than coal or oil.

What about internal combustion? The first artillerist, whoever he may have been, invented the process. His cannon was a cylinder, his cannon ball a piston, and the gas produced by burning gunpowder his working fluid. Of course, as I noted earlier, this single-cycle engine threw away its piston every time it was fired, and the artillerist had no wish to do anything else.

In the era that followed the invention of the atmospheric

Three early steam locomotives: Top, "Catch Me Who Can," the single-cylinder engine of 1808 (*a*), was built by Richard Trevithick and, drawing a single carriage loaded with paying passengers, ran around a circular track in a London park. "Puffing Billy," at center (*b*), built in 1813 by William Hedley, was a two-cylinder engine. At first mounted on four wheels, as seen here, it damaged the fragile rails. It was converted to an eight-wheeler in 1815 and ran that way until 1830. Bottom (*c*) is George Stephenson's first Killingsworth engine, a two-cylinder model, built in 1815. Note chain-and-sprocket drive. The Killingsworth engine weighed about six tons.

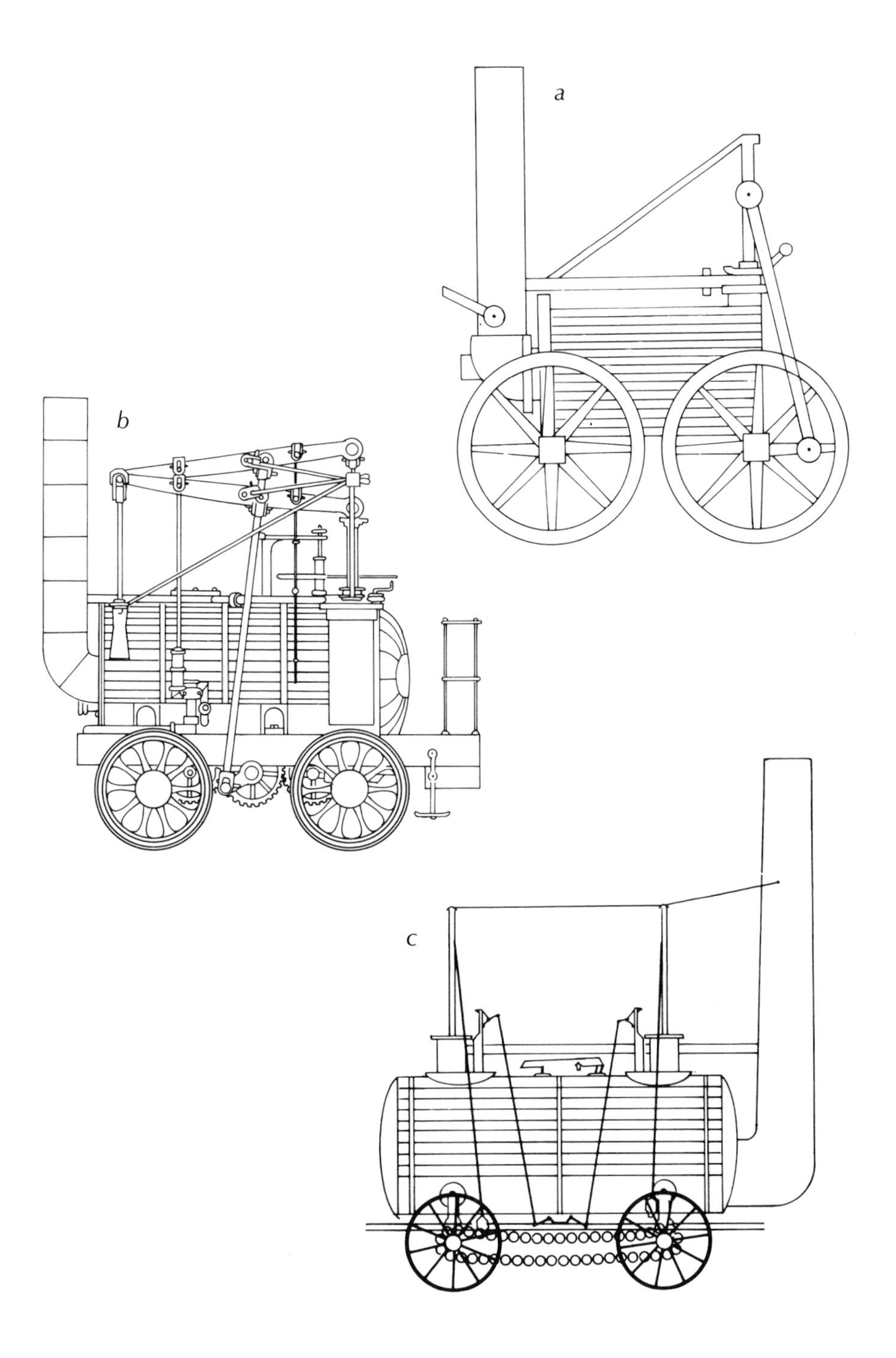
a
b
c

engine and Watt's perfection of steam power, however, a number of thoughtful men began to meditate on internal combustion. Among them was Nicolaus August Otto, a German salesman (he traveled the Rhineland in the 1860s on behalf of a wholesale grocer) and self-taught engineer. Otto decided to improve on the artillerist's cannon and use the same cannon ball over and over again. He started by pointing the cannon straight up and completely redesigning the cannon ball. He loaded his cannon not with a sphere but with a long metal ramrod that was stabilized in its upward flight by a right-angle fitting with a pair of sleeves affixed to its top end. The sleeves slid along two fixed guide rods. Instead of gunpowder, Otto used a mixture of illuminating gas and air that, when ignited, became his working fluid. Admitting just the right amount of the explosive mixture into the breech of his upright cannon, he set the gases on fire. The great expansion of the burning gases, in turn, drove the ramrod up the cannon barrel — but drove it only so far.

This was the essential art of Otto's upright engine: to drive the rod-shaped piston *almost* out of the cannon-shaped cylinder. The rod carried a strip of gear teeth on one side. A clutch kept the teeth disengaged from a drive wheel during the piston's upward flight, but once at the top of its rise the rod's array of teeth locked onto the wheel and its downward fall made the drive wheel rotate. Some of the force of the downward fall came from the action of gravity on the heavy piston. Some was atmospheric, the result of the contraction of the burned gases as they cooled, thus forming a partial vacuum in the cylinder.

Thus in one respect, and this one respect only, Otto's internal-combustion engine resembled Newcomen's 150-year-older external-combustion engine: The power stroke was the downstroke, that is, the descent of the piston, and atmospheric pressure assisted the downstroke. But the upstroke of Newcomen's piston was produced by gravity only, as a counterweight descended, whereas Otto's upstroke involved a direct harnessing of a novel source of power: internal combustion.

What made Otto's upright cannon a practical machine was the presence of a large flywheel, integrated with the gear

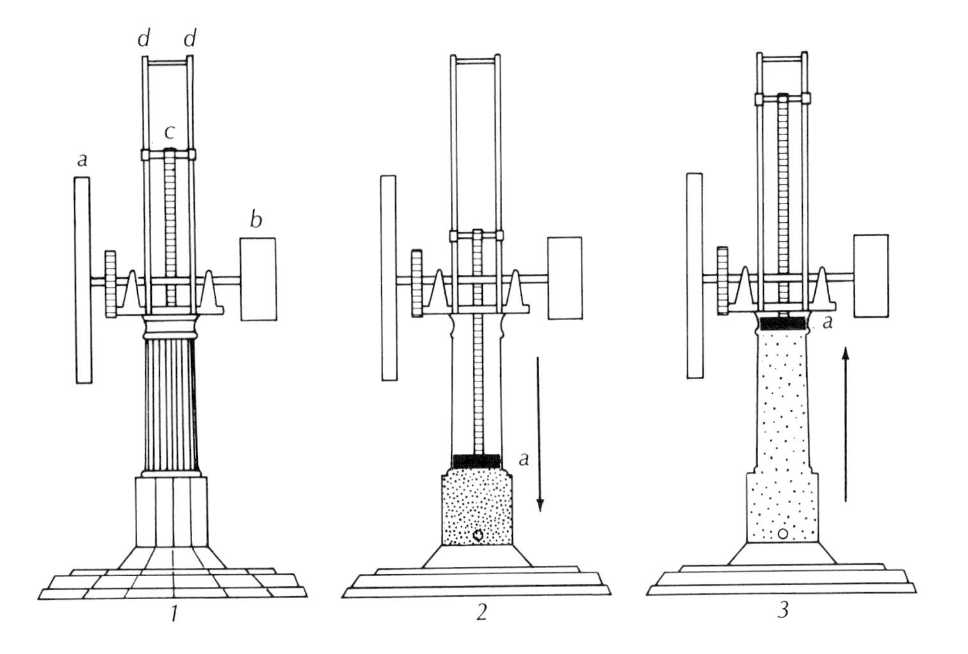

This atmospheric engine used internal combustion to create a partial vacuum. It is the upright cannon invented by Nicolaus Otto. Its cannon ball, an elongated piston, never left the barrel, and its charge was not gunpowder but a mixture of illuminating gas and air. The pedestal incorporating the barrel and breach of the cannon (1) supported an axle that had a flywheel at one end (*a*) and a pulley for a belt drive at the other (*b*). The piston, a long thin rod with teeth, widened at the base to fit the cannon snugly. It protruded from the cannon's mouth (*c*); its up-and-down motions were guided by paired rods (*d*). Here the charge has been fired and the toothed piston has begun to move back down the cannon barrel; in addition to its own weight the partial vacuum within the cannon is responsible for the piston's descent. A gear train (not shown) converts the linear movement of the piston into rotary motion. A cutaway view (2) shows the piston at the bottom of its stroke (*a*) just before the explosive mixture is fired. A second cutaway (3) shows the piston at the top of its stroke (*a*); during its ascent a clutch keeps the teeth of the rod disengaged from the drive-gears. The largest of the 5,000 Otto cannon engines built in the 19th century developed three horsepower.

train in such a way that it rotated several times per downstroke and kept on rotating when the gear train disengaged from the piston during the explosive upstroke. This maintenance of momentum gave stability to the power supply, taken off a smaller pulley wheel by means of a belt drive. Otto and his financial supporter, Eugen Langen (whose role was much like that of Boulton's with Watt), patented the upright cannon in 1866 and some 5,000 of the Otto & Langen engines were sold during the next decade. The largest of them developed three horsepower and needed more than 10 feet of clear space overhead to accommodate the upstroke of the ramrod piston.

What made Otto's upright cannon a successful machine was its smallness and simplicity when compared to even a miniature steam plant. Large enterprises like mines and big factories could accommodate a steam plant with its separate fuel depot, boiler, and engine room. The small manufacturer or craftsman, however, either stayed with water power or was dependent on animal-driven prime movers such as treadmills. The adaption of urban gas illumination changed this pattern. By the 1820s most European cities housed plants where illuminating gas was produced by the heat treatment of coal (hence the term, scarcely heard today: coal gas). Not only street illumination but interior lighting, mercantile and domestic, was achieved by piping a supply of coal gas from central storage tanks to the point of use, just as is still done with today's natural-gas stoves.

The development of a gas-burning power plant, therefore, allowed small operators to give up their riverbank locations and send their animals to pasture (or, more likely, to the knacker) and relocate anywhere they wished within the limits of the city gas supply. As the record shows, the appeal of this kind of independent power supply attracted buyers of the Otto & Langen coal-gas engines by the thousands.

Today Nicolaus Otto is best remembered for perfection of the Otto cycle. That four-stroke (intake, compression, explosion, and exhaust) internal-combustion cycle continues to run most of the 10,000,000-odd automobile and other internal-combustion engines built round the world each year. Such were the times and so many the ingenious tinkerers that it was actually a French experimenter, Alphonse Beau de Rochas, who conceived and patented the four-stroke cycle in 1862, while Otto was still hawking groceries and drawing plans. Beau de Rochas, however, never went on to develop a practical four-stroke engine and the patent lapsed. When Otto incorporated the four-stroke principle in a horizontal engine called the "Silent Otto" in 1876, the cycle soon became known as his rather than the French inventor's.

The Silent Otto, like the upright cannon, burned coal gas and developed about three horsepower. Its crankshaft rotated at 180 revolutions per minute, however, which was many times faster than the main shaft of the upright cannon. In the end, more than 35,000 Silent Ottos were bought and set to work worldwide as individual power plants. Whether Otto himself envisioned the use of his far-from-nimble prime mover as an automotive power source before his death in 1891 is uncertain. Two of his later associates — Gottlieb Daimler and Wilhelm Maybach — were destined to be pioneers of automobile and aircraft engines. These smaller and far more powerful prime movers were almost unthinkable, however, until the use of petroleum vapor as a working fluid was made possible by Daimler's invention of the carburetor in 1885. In any event, the step from gas to gasoline as a working fluid came eventually and, with it, the step from stationary to mobile internal-combustion engines. Their first use was in automobiles but aircraft engines were soon to follow.

Now it is time to sniff out another branch along our trail. It leads us to the discovery and development of a unique phenomenon and one that has already proved to be mankind's most important single source of power: electromagnetic energy.

CHAPTER
11

The Phantom Forces

WHEN WE began this journey together — tracing man's search for sources of energy, and his manipulation of various kinds of power, beginning with his own muscle — I pointed out that our planet and all who dwell here are ultimately dependent on solar energy. Sunlight is the driving force in photosynthesis, the starting point of the food chain on which we all depend. The fossil fuels that we are now beginning to exhaust are the leftovers from millions of years of accumulated solar energy. The water that surges through our hydroelectric turbines derives its kinetic energy from the solar cycle of evaporation and rainfall. Even the fission elements that make steam in our nuclear reactors are ultimately leftovers from the enormous pressure cooker that was our primordial solar fireball before this planet ever existed.

To bring on stage man's most recently developed and previously unrealized source of energy I want to present another way of looking at our sun-driven planet. We must begin by examining the four basic forces in nature. These four forces are present not only on earth and throughout our solar system but also, as far as we are able to measure them, everywhere in the universe. As we approach the end of this narrative one of the four forces is our particular concern.

The first (and weakest) of the four basic forces is the force of gravity. It is this force that orders the movement of the planets of our solar system around the sun and the movement of satellites around the planets that possess them and, more important from our day-to-day point of view, lets all of us distinguish up from down.

The second basic force in nature is called the "weak" force. It operates at the level of the individual atom and is thus of particular interest to physicists. Other than acknowledging its part in the fusion processes of stars, the sun included, here we need only tip our hats politely and pass on to the third force.

This is the electromagnetic force. As its name makes plain, this is the force responsible for the phenomena respectively called electricity and magnetism that we observe in the fantastically wide range of radiant energy known as the electromagnetic spectrum. Not too many decades ago the only part of the electromagnetic spectrum most men recognized was a very narrow band of visible light and additional sidebands in the wavelengths of infrared and ultraviolet light. It is the energy at these visible and near-visible wavelengths, endlessly showered upon us by the sun, that pumps water vapor and causes plants to grow. Today we know how to produce radiant energy artificially that is identical both with these and with many other wavelengths of the electromagnetic spectrum. Among the familiar examples are x-ray machines (one of the earliest third-force energy radiators to be perfected), radio and television transmitters and radar. What is less well-known about the electromagnetic force outside the realm of the exact sciences is that it is also responsible for the chemical interactions between molecules.

The fourth and last of the four basic forces is also the strongest, so it is called, logically enough, the "strong" force. Like the weak force, it operates at the level of the atom. Specifically, it binds protons — the individual charged particles that are present in the nuclei of all atoms — to one another, even though the positive charge that all protons carry should drive these particles apart. As with the weak force, all we will do here is greet the strong force politely and pass by.

Neither need we, in returning to the third basic force in nature, waste much time by retelling the tales of man's early discoveries of electricity and magnetism. Everyone knows about compasses, our earliest devices for exploiting the paltry magnetic field that is somehow generated by the spin of our planet. Almost everyone knows that the word electricity comes from the Greek name for amber *(elektron)*, because if one rubs amber (or, better still, sealing wax) with a cloth or a bit of fur one can generate so-called "static" electricity and make sparks fly. We will pass by Leyden jars and galvanic cells, nod politely once more to Ben Franklin and his kite, and proceed directly to 1831, the year when, after four earlier failures, an English experimenter named Faraday proved that the two parts of what we now know to be nature's third force are indeed connected.

Half of this proof had been furnished 11 years earlier by a Danish physicist, Hans Christian Oersted, who demonstrated that an electric current could generate a magnetic field. Oersted placed a wire parallel to a compass needle at rest (that is, pointing north/south). Using a chemical storage battery, a recently invented novelty, as his source of electricity, Oersted then let an electric current run through the wire. When he did so the needle turned from its normal orientation to an east-west position — that is, a position at a right angle to the wire. When he cut off the current the needle returned to its normal orientation.

If an electric current can produce magnetism, why shouldn't a magnet produce electricity? This is a question we should all be glad someone as ingenious as Michael Faraday thought to ask. When he posed it to himself, although he could scarcely have been aware of its long-term implications, Faraday had begun to erect the very foundation of modern power technology. Who was this 40-year-old Englishman and what were his credentials?

Michael Faraday, son of a blacksmith, was born in 1791. Having reached working age without any particular influences on his development other than the strong religious views of his family, he was set to learn a trade. In young Faraday's case this meant being apprenticed to a London bookbinder. To bind books is likely to be an uneventful enough career, although sufficiently bookish to give a youth ample opportunity for self-instruction.

As it happened, certain leading figures of the Royal Society (England's "Academy of Science") at the end of the 18th century had founded in London an institution with the objective of teaching "the application of science to the useful purposes of life." Known as the Royal Institution of Great Britain, it incorporated foundations that supported three professorships, a laboratory, and the cost of publishing a journal. Its chief teaching method, however, consisted of public lectures and in 1812, at age 21, Faraday was taken to hear a series of lectures given by a pioneering English chemist, Humphry Davy, one of the three Institution professors and director of its laboratory.

Shortly thereafter Faraday was introduced to the great Davy and the two struck it off to the extent that in 1813, on Davy's recommendation, Faraday was appointed his laboratory assistant. Sixteen years later, in 1829, four years before Davy's death, Faraday succeeded him as director of the Institution laboratory. As might be expected, Faraday's work under Davy was mainly in the field of chemistry, but science in those days was far less compartmentalized than it is now. Davy himself, for example, performed any number of chemical experiments that utilized electrical current. (One of these, the separation of pure metal from a molten mix by passing a current through the crucible, later became the foundation of today's aluminum industry.) In any event, Faraday became more and more concerned with electrical theory and so it was that Oersted's experiment in 1820, nine years before Faraday took over Davy's laboratory, captured his attention.

How could a magnet be made to produce electricity? Beginning on August 29th, 1831, Faraday successively demonstrated four methods for producing brief surges of electric current by means of magnets. Before 10 days had passed he devised an apparatus that produced a continuous flow of current; he called it a "new electrical machine." It was, in fact, the first electric generator.

Its basis was a powerful permanent magnet bent into the familiar "horseshoe" conformation so that the magnet's two poles were separated by a narrow air gap. Mounting a circular copper disk on an axle, Faraday fitted the disk into the air gap between the two poles of the magnet so that about one quarter of the disk lay between the arms of the horseshoe. To measure what happened when he rotated the disk he used a current-detecting instrument (called a galvanometer, after Luigi Galvani, who in 1792 first described the generation of electricity by chemical action). One lead from the instrument Faraday connected to

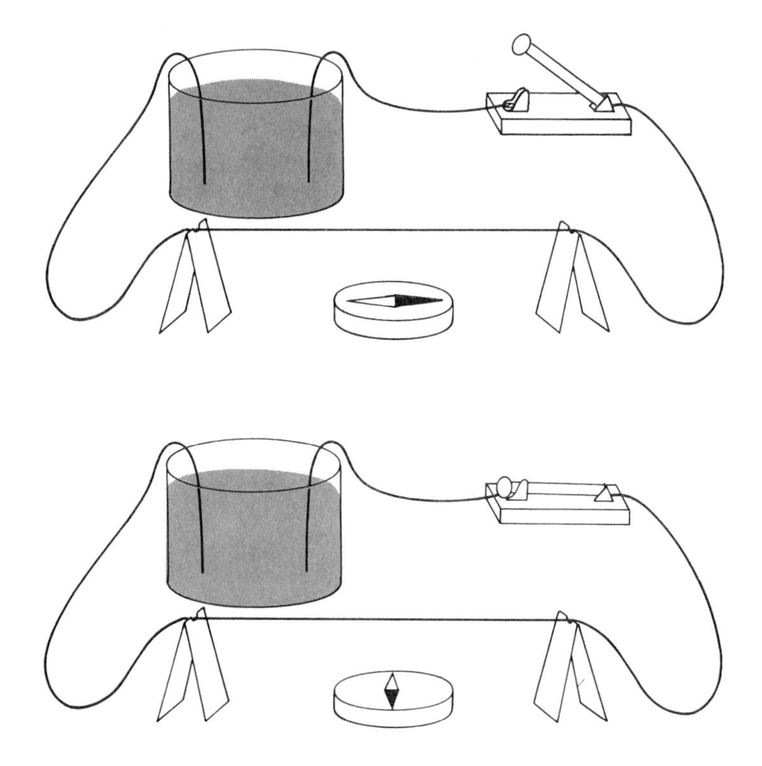

The relation between electricity and magnetism was demonstrated first in 1820 by the Danish physicist Hans Christian Oersted in an experiment like the one illustrated here. Oersted placed one part of a wire loop just above and parallel to the needle of a north-pointing compass. When he closed a switch, so that current from a battery passed through the wire, the magnetic field generated by the current deflected the compass needle so that it pointed east-west instead.

a strip of conductive metal that rested on the edge of the copper disk; today this would be called a brush. The other lead ran to the axle that allowed the disk to rotate; the axle was also made of conductive metal. When Faraday now rotated the disk, the galvanometer needle showed that the disk's movement between the poles of the magnet induced a steady flow of electric current.

I would like to be able to report that Faraday promptly petitioned Parliament for a patent and then, perhaps under the name of Consolidated Faraday, built hundreds of these generators and electrified England. Alas, on the contrary, once having published his findings Faraday pressed no further. History records, instead, that the first commercial use of the principle of his "new electrical machine" was trivial. Faraday generators were built by others in order to send weak electric currents through the bodies of various patients whose physicians, accepting a

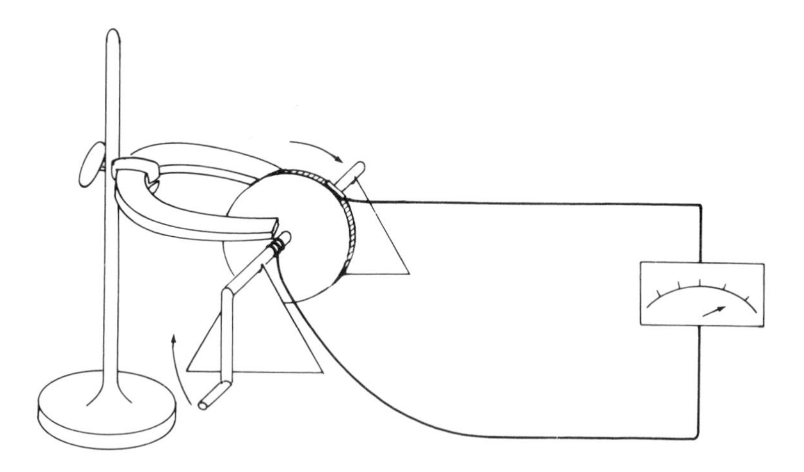

Magnetic generation of electrical current: Michael Faraday's key experiment of September 1831, is illustrated here. When a copper disk is placed between the poles of a horseshoe magnet and then spun, its continuing reaction with the invisible magnetic lines of force generates a current. The wire loop connecting disk edge and axle to a galvanometer, at right, carries the current to the instrument and the needle of the galvanometer is deflected.

medical fad of the period, thought this precursor of both shock therapy and electrocution would promote health.

When I say that Faraday pressed no further I refer only to the "new electrical machine." Between 1831 and 1851 this great explorer of a new realm published no fewer than three volumes devoted to his experimental findings. A concluding volume appeared in 1855; it summarized all that was then known about magnetism in general and electromagnetism in particular. So far as the unification of theory is concerned, perhaps the most interesting of all of Faraday's studies was one that allowed him to rotate the plane of a beam of polarized light by applying a magnetic field. Thereby Faraday established an observable connection between two then seemingly unrelated phenomena: light and electromagnetism.

This is no place to trace the development of the electric generator step by step. Let it merely be noted that the first patent for a generator using electromagnets rather than permanent magnets, as Faraday's machine did, was granted in 1863, four years before Faraday's death. The first standard dynamo was perfected and marketed in 1870, three years after. It was the 1870 dynamo (short for "dynamoelectric machine") that soon did such electric generation as the time required, for example, powering the widely used process of electroplating (a cold-bath variation on Davy's method of capturing pure metal from a solution) and providing the electric current for the brilliant carbon arcs that made arc lamps standard fixtures in lighthouses.

You will remember Faraday's first fortunate question: If an electric current can produce magnetism, why shouldn't a magnet produce electricity? We are doubly fortunate that, after Faraday produced his "new electrical machine," someone asked a similary fortunate second question: "If I can turn a crank and generate electricity, why can't I arrange for electricity to turn a crank?"

Omitting the crank from both ends of this question, we find, astoundingly enough, that the questioning someone was none other than Faraday himself. Moreover, he had proved to his satisfaction the power of electricity to make things move in 1821, or a decade before his proof that moving something in a magnetic field would generate electricity. Moreover, a fellow scholar in London, Peter Barlow, soon thereafter produced the prototype electric motor. It was necessarily battery-operated because dynamos would not be developed for sometime to come, but Barlow's "wheel" is still used today to power, for example, the air compressors for home aquariums and to turn the pointers on your electric meter.

What Barlow did was to suspend a copper disk, free to rotate, between the poles of a horseshoe magnet. When he then let electric current flow from the center of the disk to its circumference, the disk rotated. So, there you are; the first electric motor was devised as a laboratory curiosity before the invention of the dynamos that might have supplied such motors with enough power to make them useful.

We must now let the curtain descend while I give as brief a lecture as possible on the subject of A.C. and D.C., abbreviations known, often in a non-electrical context, to every schoolboy and schoolgirl. The first abbreviation stands for "alternating current" and the second for "direct current." When everyone was developing dynamos in the mid-1800s, the current being generated was either D.C. or A.C., depending on the design of the generator. At first, D.C. dynamos predominated, but a number of considerations — in particular the greater efficiency of high-voltage A.C. power line transmission — gradually brought about a decline in the generation of direct current. A factor of equal importance is that A.C. can easily be turned into D.C. ("rectified") so as to supply direct current to the many electrical devices that demand it (radio and television receivers, for example) today. It is , however, a nuisance to convert D.C. into A.C.

The first really practical motors were produced by a French electrical engineer with the mellifluous name of Zénobe Théophile Gramme, who had been a very important contributor to the development of dynamos in the latter 1800s. Gramme and a colleague unveiled a combination dynamo-motor — a reversible pair of regular Gramme generators — in 1873. Because all of his dynamos produced direct current, Gramme's motor was also a D.C. device.

In less than a decade, A.C. dynamos had begun to compete with D.C. dynamos and so a market for A.C. electric motors developed. The need was met some five years later, in 1888, by an engineer from imperial Austria: the Croatian, Nikola Tesla. An emigrant to the U.S. in 1884 at the age of 28, Tesla had a long career in electrical engineering and electromagnetic theory. He worked with two of America's electrical pioneers: with Thomas Edison (briefly) and with George Westinghouse (extensively). The A.C. motor he invented four years after his arrival in America was manufactured by Westinghouse; in the years that followed, any number of improvements were added by others in Switzerland, Germany, France, and England. The basic invention, however, was his, and once direct-current dynamos began to disappear in the decades following 1900, A.C. systems naturally filled the gap, in addition to being applied to innumerable tasks in their own right.

Ninety years ago the most common of these tasks was propelling track vehicles: the streetcars and suburban trolleys that have now almost disappeared from America. (Notable exceptions are the excellent streetcars of Toronto and the 15-mile trolley line between San Diego and the Mexican border.) Even that once commonplace example of a working union between internal-combustion engine, electric generator, and electric motor — the diesel-electric locomotive — is known today only to the ever-diminishing number of those who see the nation's railways in action. But these transport uses, like the more direct application of electricity to lighting and heating and to industrial chemistry, are almost trivial compared to the other uses of electric motors.

Think about this. In the days before electric motors, it was possible to work in a factory lit by arc lights (and later by light-

bulbs) but the power available at the workbench was still rotary motion (brought off a main shaft, as often as not, by a drive belt), and what made the main shaft go round was some external-combustion or internal-combustion engine or, if even more old-

Steam-powered factory: in the days before the electric motor and the internal-combustion engine, machine-shop and similar opera-

tions were possible only in places where a water wheel, providing main-shaft rotary motion, could drive arrays of pulley belts that, in turn, drove such machine tools as lathes. After steam engines came into use, factories no longer had to be built by millponds. This is I. M. Singer's sewing-machine factory, at Centre and Elm streets in New York; the year is 1853. All the machine tools seen here were used to finish rough castings. They took their belt power from a single main shaft kept in motion by a steam engine.

fashioned, a water wheel. With the advent of electric dynamos and electric motors, however, rotary motion became available everywhere! On your wall, where the electric clock counts the passing seconds. On your workbench at the flick of a motor switch. At the turntable of your record player. At your desk, if you use an electric typewriter. In your hands when you hold a power drill or a power saw.

Your power plant may be miles or scores of miles away but you need no more intimate connection with this primary energy converter than your own wall plug and a length of wire. You can vacuum your rug, dry your hair, and power the compressors that make your ice, keep your food cold and your room cool in summer, all by plugging in. (Rotary motion to one side, you can also cook your food, warm your room, and press your clothes by plugging in, but stove, heater, and iron are not motor devices. They are hungry watt-gobbling resistance devices and any sensible person today uses them as little as possible.)

On balance, then, thanks to Faraday, Gramme, Tesla, and uncountable scores of others, you now have virtually unlimited power at your elbow or in your hands. This power, moreover, is the product of a kind of energy that was scarcely known and wholly unexploited 300 years ago.

So far I have directed your attention primarily to the gen-

Steam into electricity: this composite illustration, which appeared in the issue of *Scientific American* published on June 13, 1890, shows the fifth plant for the generation of electricity built in the New York area. It is the Edison Electric Illuminating Company's central station on Pearl Street in Brooklyn. Its 14 direct-current dynamos were belt-driven by seven coal-fired steam engines running at a rate of 220 revolutions per minute. Lighting was the plant's main business; its output was enough to illuminate 21,000 light bulbs, but some electric power was already being used to drive factory motors. By this time Thomas Edison dominated electric-power generation in the U.S. His companies operated some 400 plants as compared to fewer than 100 operated by his largest competitor.

SCIENTIFIC AMERICAN N.Y.

eration of electricity and its power applications. It is time now to examine other, even less credible, aspects of electromagnetism.

A decade before Michael Faraday died in England a baby boy was born in Hamburg, across the North Sea. In his short life he was to open up another potential function of electromagnetism, one that is quite as ever-present in your life today as are the electric generator and motor. The infant was named Heinrich Rudolf Hertz, and the electromagnetic rabbit he pulled out of his hat at age 30 was a practical demonstration (using a far from complex laboratory apparatus) that a spark discharge caused electromagnetic waves to radiate out through space in all directions. Technically speaking, what he did was to show that the discharge of what is known as a condenser is oscillatory; the stored electric current does not simply drain away but surges rapidly back and forth, losing a little of the charge with each oscillation. Each such oscillation, in turn, radiates an electromagnetic wave.

For the "sending" side (or "radiator") of his apparatus, Hertz used two brass rods, each about a foot long. At one end of the rods was mounted a square zinc plate about two and a half inches to a side, and at the other was a polished brass ball. Each rod was supported by an upright stand so that the two balls were separated only by a narrow air gap; in turn, both rods were connected to an induction coil. In this condition Hertz's radiator was the equivalent of an uncharged condenser.

Hertz then charged his condenser by feeding current from a battery into the induction coil. The next step was to make fine adjustments in the distance between the two brass balls. At the right spacing a lively spark jumped across the gap, as the condenser discharged in a single pulse of very high voltage, propagating an oscillatory wave of entirely invisible and intangible electromagnetic energy.

How could such waves be detected? Hertz had devised just the apparatus for the job. Called a resonator because it was tuned to the resonant frequency of the radiator, it was a simple

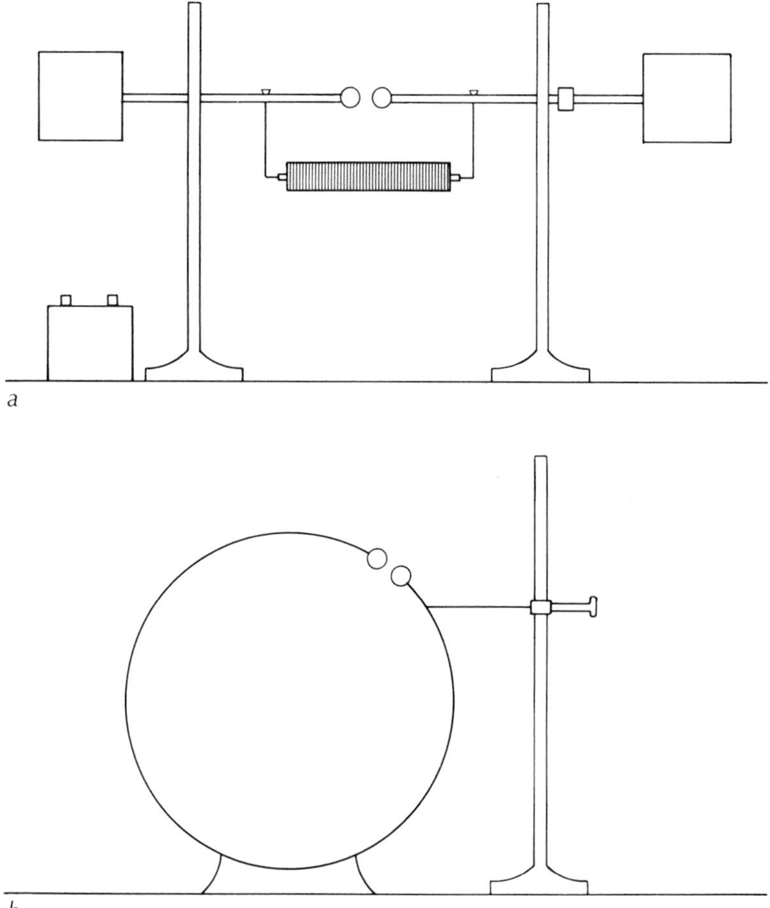

Hertz "radiator" (*a*) and "resonator" (*b*), or what we would call a sender and receiver today. First the battery was connected to an induction coil (*below the spark gap*) and the coil was charged. The battery was then disconnected and the gap was narrowed until a spark jumped across it, propagating an electromagnetic wave that traveled in all directions. Since the wave was invisible and intangible, how did Hertz detect it? With his resonator. When the gap between the balls at each end of this nearly closed wire loop was properly adjusted and Hertz's radiator was then discharged, enough of the energy it broadcast was collected by the loop to make a small spark jump across the narrow gap between the brass balls.

ring of wire, about a foot in diameter, mounted on insulating supports. The ring was not complete; there was a gap at the top where a pair of brass balls were narrowly separated. A screw device allowed Hertz to alter the size of the air gap.

When the resonator (or detector) was properly positioned and the radiator was fired, another spark, far fainter than the radiator's, appeared between the brass balls of the detector. The invisible and intangible electromagnetic wave had traversed the space between radiator and detector, delivering enough energy to produce a spark!

Over the next two years, using apparatus not much more elaborate than simple radiators and detectors (soon they would be called transmitters and receivers), Hertz went on to measure the length of the waves and to show that with respect to such properties as wave form, reflection, refraction, and polarization they exactly coresponded to waves of light. Here then, within decades of Faraday's proof of electromagnetic polarization, Hertz had demonstrated that the electromagnetic spectrum is all of one piece. In terms of wavelength, Hertz's radiator was broadcasting "short" waves in that part of the spectrum that stretches far, far out beyond infrared, where wavelengths are measured in meters (short-wave radio) and kilometers (long-wave radio); our standard broadcast band lies on both sides of a wave one kilometer (five-eighths of a mile) long.

In September 1889, Hertz lectured the German Society for the Advancement of Natural Science and Medicine at a meeting in Heidelberg on the relations between light and electricity. Less than five years later he was dead at the age of 37.

Hertz's short waves (at first one-sixth of a meter and later six meters in length) were detectable at a distance of a few yards. By 1890 the French had extended the broadcast range to a few tens of yards, and by 1895 British experimenters had stretched it to half a mile and three-quarters of a mile. More was soon to follow.

Hertz was born a decade before Faraday's death. Two

decades before Hertz's untimely death in 1894 an infant boy in Italy received the mouth-filling name of Guglielmo Marconi. Had Hertz lived longer, Marconi might have been his pupil. As it was, in the year after Hertz died the young Italian student of physics, using homemade apparatus, managed to send and receive Hertzian waves farther than the English had done that same year. His detector intercepted the signal from his radiator at a distance of more than a mile.

When he found that a Hertzian radiator could be operated in an intermittent manner so that a sequence of broadcast waves could duplicate a telegraphic code (for example, the code devised by Samuel Morse), Marconi recognized the great advantage of wireless telegraphy. He sought the financial assistance of the Italian government for his experiments. He did not receive it.

Marconi's mother was a Jameson and so he had grown up fluent in English and also was able to make use of her family's English connections. Visiting England, he found that the enthusiasm for wireless telegraphy was far greater there than in Italy. He soon received backing for the first system of ship-to-shore wireless communication. He set up a sending and receiving apparatus at the lighthouse at South Foreland, just east of Dover, in Kent. Duplicate apparatus was installed in the lightship guarding a notorious hazard to navigation in that coastal area, the Goodwin Sands. The installation was completed in December 1898 and lightship and lighthouse remained continually in wireless contact (by Morse code) for the next twelve months.

This was persuasive proof of the reliability of Marconi's wireless and he followed it with another: cross-Channel wireless. His English installation was also at South Foreland and its French counterpart was at Wimereux, near Boulogne, 30 miles away as the crow flies. This international wireless link was quite as successful as the ship-to-shore system had been. In July and August of 1899 two cruisers of the British royal fleet — Her Majesty's Ships *Juno* and *Europa* — maintained contact during fleet maneuvers by means of Marconi wireless telegraphy, thus proving its value in ship-to-ship communications.

Marconi's successes were in large part due to simple improvements in Hertz's laboratory-scale radiators and detectors. To increase range, he lowered the frequency of his radiators' emissions, so that far longer wavelengths were broadcast. The radiators were made vastly more powerful by using steam-driven electric generators instead of batteries and by providing them with an extensive antenna system (as the lower frequencies demanded) and a connection to the ground.

The detectors were rendered far more sensitive by the addition of a "coherer," a device pioneered in France and improved in England. The coherer was, in effect, a triggering device. Instead of the simple wire ring used by Hertz, the receiver antenna fed its tiny signal into a circuit that was battery-powered and that activated a standard telegraphic dot-and-dash relay. The current from the battery could not flow through this circuit and activate the relay under ordinary circumstances because of the presence of the coherer.

What prevented the flow of battery current through the coherer was a pinch of fine powdered metal inside it. In their normal state the particles of metal were not in close enough contact with one another to allow the flow of battery current; they did not "cohere." When, however, the arrival of a Hertzian wave at the antenna made a tiny trickle of oscillating current pass through the coherer, this caused the powder, true to its name, to cohere, and the circuit between the battery and the telegraphic relay was completed. Click — "dot." To prepare the coherer to react to the arrival of the next Hertzian wave it was only necessary to tap the coherer tube and thereby render the powder incoherent once more. Sounds like Rube Goldberg? Maybe, but it worked.

Ship-to-shore, cross-Channel, ship-to-ship: what next? Why not transatlantic? By early December 1901, Marconi had a 400-wire transmitting antenna built in southern Cornwall, draped on four 215-foot lattice towers spaced 200 feet apart. Instead of using the familiar spark-gap discharge with its pair of copper balls, the condensers of his transmitter (charged by high-tension transformers tied into a steam-driven electric generator) were discharged — and the waves propagated — by a disk rotating at high speed.

Marconi was already waiting on the other side of the Atlantic, near St. John's, Newfoundland, where he had set up a receiving apparatus. His colleagues in Cornwall had been instructed to come on the air at regular intervals, repeating a single signal. This was the three dots representing the letter "S" in Morse code (two-thirds of the later international distress signal: three dots, three dashes, three dots, or S O S).

Pause with me for a moment now, and tip your hats once more to old Ben Franklin. If that 18th-century kite-flier could have been with Marconi on December 12, 1901, it would surely have made his heart glad to see that the pioneer in telecommunications had rigged the simplest of antennas to intercept the signal from Cornwall. It consisted of 400 feet of copper wire lifted high in the air by a box kite. And on that day it did more yeoman work than any of old Ben's kite strings ever had, successfully carrying the three-dot signal from Cornwall, 3,000 miles away, down to the coherer in Newfoundland.

Why do we call radios "the radio" while many in England call them "the wireless"? Because the device that Marconi proved to be capable of transatlantic communications was named the wireless telegraph in England and the radiotelegraph in America ("radio" had come into use sometime before the 1900s in both England and America as a prefix signifying radiation in such words as "radiometer" and "radioactive"). We clipped "radiotelegraph" down to "radio" and the English likewise dropped "telegraph" from their expression, leaving an adjective to do the work of a noun. ("Wireless" is falling out of favor, now, and "radio" is taking its place.) When "sight at a distance," or television, first appeared a few decades ago, both parties did much the same thing. The English clipped the word down to "tele," pronounced Telly, and we broke the word in half and used only their two initials to coin the new abbreviation, "T.V."

In any event, we need go only a little beyond December 1901 to see the birth of voice radio and other examples of man's utilization of the electromagnetic spectrum. Marconi's transatlantic success narrowly preceded Christmas, 1901. It was actually on Christmas Eve, five years later, that two kinds of man-made sound — voice and music — were sent far through space on Hertzian waves.

Before this could be done it was first necessary to produce Hertzian waves in a nearly continuous fashion, rather than in short bursts. The great English experimenter John Ambrose Fleming, writing early in the 1900s about the evolution of wireless telegraphy, described it this way. When condensers are used to provide the oscillation ". . . the intervals of silence are nearly 400 times as long as the intervals of activity. The condenser method . . . is analogous to the production of air vibrations by twanging a harp string at short intervals. What is required, however, is something analogous to an organ pipe which produces a continuous sound."

Such analogues to organ pipes were soon invented in a variety of forms. At first their use was confined to radiotelegraphy. The nearly continuous stream of waves was chopped into pieces at the transmission end and detected as coded bits at the receiving end. Then some of the rapidly growing technology related to the telephone was added; an earphone replaced the receiving relay and the chopped signal became audible in a new way. Marconi's clicked "dots" became noisy "dits" and the distress signal was turned into "dit dit dit dah dah dah dit dit dit." Or, in a fast fist, "divididit dadadah dididit."

It was at this point that a Canada-born scientist and inventor, Reginald Aubrey Fessenden, substituted a microphone for the telegrapher's key at the sending end. As early as 1904 he was offering wireless telephone senders and receivers for sale and they were popular for ship-to-shore use. His pioneering work in the field set the stage for Christmas Eve, 1906.

"Reg" Fessenden's firm, the National Electric Signaling Company, ran a private ship-to-shore wireless service that included a link between Brant Rock on the Massachusetts shore and the town of Plymouth, some 11 miles away. Fessenden had converted this link into a wireless telephone system. Fishermen of the South Shore fleet had already on at least one occasion called to the Brant Rock station, asking for current prices on the Boston fish market, and by December 11 the time seemed right to demonstrate how far the Brant Rock voice transmitter could broadcast. Prominent witnesses were invited to visit the station on Christmas Eve and a few days before the demonstration, using conventional wireless telegraph, Fessenden notified various naval vessels and United Fruit Company ships equipped with both telegraph and telephone receivers of the impending demonstration.

The Christmas Eve broadcast began with a brief speech read by Fessenden, announcing the demonstration. This was followed by the playing of a phonograph record of Handel's "Largo" and a live violin rendition of Gounod's "O, Holy Night" (Fessenden doubled as violinist). Next Fessenden read a Biblical quotation appropriate for the occasion: "Glory to God in the highest, and on earth peace, good will toward men." Brant Rock signed off by wishing its listeners a Merry Christmas and inviting them to hear another broadcast on New Year's Eve.

Ships at sea as far south as Norfolk, Virginia, picked up the Christmas Eve broadcast and Fessenden received acknowledgment from listeners in the West Indies following his second program. Not long after this, having installed a more powerful alternator at Brant Rock, he was astonished and delighted to learn that routine conversations with Plymouth were being overheard in Scotland. The air waves have never been entirely voiceless since.

Essentially, all that separates 1906 from today and tomorrow in these ranges of the electromagnetic spectrum is the absence of a device for grasping the faint far-flung Hertzian waves and amplifying them efficiently. You who are wise in the ways of electronics will have already recognized in the powdered-metal "coherer" a kind of great uncle of such a device. How more efficient devices were discovered is a rather odd story. In the simplest form it begins with Thomas Alva Edison.

In what Isaac Asimov calls Edison's "only discovery in 'pure' science," the great inventor one day added an extra element—a plate—to one of his incandescent light bulbs. This random experiment led to the discovery of the "Edison effect." When the light-producing filament of the bulb was turned on, an electric current flowed from the hot filament across the empty space to the intruding plate and thence out of the bulb. This was interesting but not obviously commercial. Edison noted the fact and shelved the experiment.

At about the same time — to be exact, in 1904 — the same Fleming who invented the analogy of harp string and organ pipe was in search of a simple way to transform alternating current to direct current. Why not the Edison package: an evacuated glass envelope, the hot filament, and a metal plate? Sure enough, when Fleming fed alternating current to his plate, a cloud of electrons passed from the hot filament to the plate during that half of each cycle during which the plate carried a "plus" charge. When the other half, or "minus," part of the cycle came along, however, nothing passed from filament to plate. Fleming had invented an electronic "valve," exactly analogous to (although immeasurably faster in operation) an hydraulic one and to those valves in the internal-combustion engine that let in the explosive gas and let out the exhaust.

You have already recognized this rectifier, with its pair of electrodes, as the common "diode" tube of radio before the days of solid-state. Three years later the next advance toward efficient amplification appeared: the "triode." Obviously, with such a name, the vacuum tube contained not two but three electrodes. This is what made it work. The third electrode, as the American engineer Lee De Forest contrived it in 1907, was located between the hot filament and the plate. This third element was called a grid, perhaps because of its griddle-like appearance, and its effect was to control the flow of electrons from filament to plate. If, as De Forest did, one fed to this grid the very weak, fluctuating Hertzian signal that reaches a radio antenna, the flow of electrons from filament to plate would fluctuate in the same way. But far from being weak, this corresponding flow can be as strong as is wished!

We call such radio tubes (and the solid-state circuits that have now almost everywhere replaced them) amplifiers. If De Forest had thought back to the Greek machines and at the same time been whimsical enough, he might have called his triode an electronic lever: it was certainly a massive multiplier of force. Indeed, so effective is the lever of electronic amplification today that the listeners at our space centers can restore to normal strength radio signals being broadcast from hundreds of millions of miles away, signals that reach earth antenna systems with no more energy than 0.000,000,000,000,000,14 w., or less than a tenth of a quintillionth of a watt.

In our examination of the electromagnetic spectrum we have looked so far at what might, in typically man-centered fashion, be called the middle of the spectrum (that is, the part we can see) and also at that part of the right side, where the Hertzian waves tumble free, their length from crest to crest ranging from fractions of a meter to tens of thousands of miles. Between these two very unequal zones lies the infrared region, where the frequencies are lower than those of visible red light and the length from crest to crest is measured first in fractions of microns and then in microns. Further to the right, approaching the radio frequencies, lies the microwave region, where the frequencies continue to drop and wavelengths are measured first in millimeters and then in centimeters.

How about the left side of the spectrum, where frequencies rise and wavelengths grow progressively shorter as visible light vanishes into the ultraviolet? Well, for one thing, the shrinking wavelengths grow so small that a novel unit of measurement is required to describe them. This is the Ångstrom, one hundred-millionth of a centimeter; it is named for the 19th-century Swedish astronomer who first proposed its use for measuring the wavelengths of light. The steps that lead to the Ångstrom are as follows. One meter equals 1,000 millimeters; one millimeter equals 1,000 micrometers (more commonly called microns); one micron equals 1,000 nanometers (still called millimicrons by some). In these three steps we have descended from one meter to one billionth of a meter and are in reaching distance of an Ångstrom, which is one tenth of a nanometer.

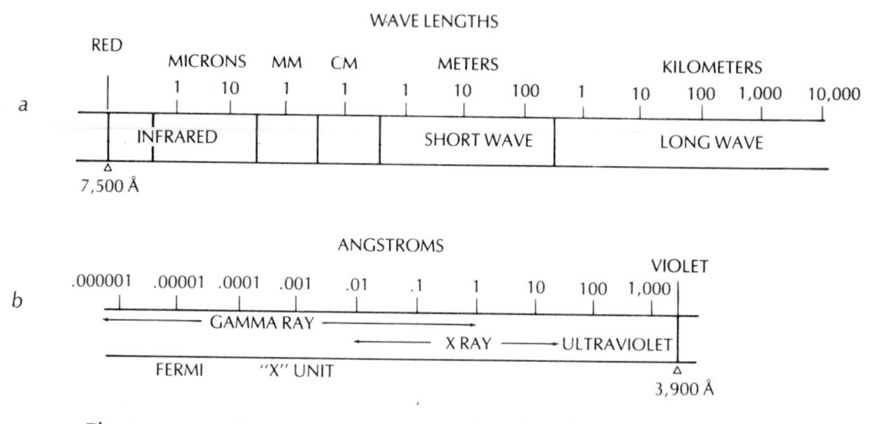

Electromagnetic spectrum to the right of the band of visible light is shown at the top (*a*) of this illustration and the spectrum to the left of that band is shown at the bottom (*b*). Scale in both is logarithmic: each interval is greater or less than the preceding one by a factor of 10 (otherwise the scale would stretch out into the back yard).

The wavelengths at the "long" (or red) end of the visible spectrum measure about 7,500 Ångstrom units, a rather more convenient expression than 0.000075 centimeter. The wavelengths at the "short" (or ultraviolet) end measure some 3,900 Å, the abbreviation for this unit. This, however, is only a beginning. As we move further left we reach the wavelengths of x rays, from below 100 Å to below one one-hundredth Å. This area of frequencies overlaps the frequency range of gamma rays and other forms of ionizing radiation that are lethal in character. "Ionizing" simply means that the radiation so characterized can zap some electrons away from an atom and thus upset its normally neutral charge: this is why x-ray technicians wear protective clothing.

From this point and out to the left beyond it still smaller units of measurement come into play. For example, one one-thousandths of an Å is known as an X unit and one one-hundredths of an X unit is a Fermi (so named in honor of the Italian refugee physicist, Enrico Fermi, a leader among those who gave America the atom bomb). We need not linger in this region of infinitesimals, where the radiation can kill you. Let us return, instead, to the home of the microwaves, beyond the infrared on the right half of the spectrum. It is in this range of frequencies that we will encounter a use for electromagnetism millions of years removed from but very much like man's earliest perfection of a cutting tool.

As a starting point, consider once more De Forest's triode amplifier. This wonderful variant on Fleming's electronic valve achieved its effect by modulating the torrent of electrons that poured off the tube's hot filament. The amplified input, however, became married to a number of stray electrons in the process and these strays produced distracting "noise." As a result, particularly when incoming signals are faint, electronic engineers make every effort to keep the ratio of signal to noise firmly biased in favor of the signal.

But what if an amplifier could be constructed that did away with the torrent of electrons altogether? The signal-to-noise ratio would jump from, say, 60/40 to 100/0: velvety silence. With this kind of noiseless amplification one could detect and bring up to normal strength signals of almost unimaginable faintness. Marvelous, but how to do it?

Well, fortunately, matter, at the level of the atom and the atomic associations we call molecules, exists in a random variety of energy "states," some lower and some higher than others.

Interactions between atoms produce changes in their state and if one atom falls from a higher to a lower level of energy it will emit a "quantum" of energy. *Quantum* (from the Latin word *quantus* that means "how much?") may have been selected in 1900 by the illustrious German physicist Max Planck as a kind of gentle joke on his colleagues. If he was poking fun at the tendency of scholars to coin Latin names for new concepts, Planck was nevertheless in deadly earnest about his proposal. This was that electromagnetic energy exists not as a seamless web but in the form of tiny discrete parcels, or "quanta," that are not all the same but vary in "size" according to the amount of energy they represent.

Atoms and molecules are changing state and emitting or absorbing quanta all the time, but the process is so chaotic and so small-scale as to be insignificant and virtually unobservable unless watched for. But what if one were able to "pump up" an aggregation of atoms to the point where an additional stimulus would trigger a cascade of "coherent" quanta, all in step and all of the same frequency? If even a very small stimulus could trigger the cascade, you would have one hell of an amplifier! Moreover, because no swarm of electrons was involved, such an amplifier would be silent as the grave: all signal and no noise. What might such a process be called? How about "amplification by stimulated emission of radiation"? That's quite a mouthful; let's instead just take the initials. A.S.E.R. Or ASER for short.

Something along those lines was on the minds of some workers at the Bell Laboratories a little more than 20 years ago. The New Jersey experimenters were able to pump up the chromium atoms in a synthetic ruby (this gemstone consists of a crystal lattice of aluminum oxide that includes chromium atoms) to a point that required only the faintest further input to trigger an amplifying cascade of coherent quanta. What uses could be found for such a noiseless amplifier?

The Bell Labs people found that the first to ask for pumped-up rubies were radio astronomers, eavesdroppers listening to the often faint natural radio emissions generated by cosmic processes. Because these signals are often short in wavelength, the final formal name for the Bell Labs ruby-amplifier process was "microwave amplification by stimulated emission of radiation," or MASER.

Today, arrays of maser antennas receive and amplify not only the chatter of the cosmos but also the painfully weak radio signals broadcast by planet-probing spacecraft. Those binary-coded pictures of the moons of Mars or, more recently, of the volcanic eruption on Io, the fourth moon of Jupiter, some 600 million miles from the earth, could never have been received without noiseless microwave amplification.

Change one letter in maser and we find that man's every-growing mastery of energy has carried us to the brink of fantasy: if not *Star Wars* at least *Space Opera.* The letter, of course, is "L," for "light," and the abbreviation LASER stands for "light amplification by stimulated emission of radiation." Laser is the last word in the title of this book. The finicky can mourn the coinage of such new words but things could be worse. In July 1960, when this kind of amplification was first achieved, the instrument that released a light beam composed of coherent quanta was known as an "optical maser." We have at least been spared that.

The very first laser used a pumped-up ruby much as the Bell Labs maser had done. A specially shaped cylindrical synthetic ruby, about an inch and a half long and a quarter of an inch in diameter, it had both ends carefully polished and then silvered like mirrors. Its designer, Theodore Maiman, pumped it up with a flash tube that produced ordinary, incoherent light. Receiving these pulses, the excited chromium atoms in the ruby lattice began to emit quanta. Many of these simply flew off at angles and vanished through the sides of the ruby cylinder. Some, however, traveled the length of the cylinder and reached one or another of the silvered ends. Hitting the mirrors, these quanta would bounce back the way they had come, hit the opposite mirror, and repeat the trip. Each time one of the bouncers had collided with a pumped-up chromium atom the collision would knock loose another quantum that might also become a bouncer.

Soon, between pumping and bouncing and bouncing and more pumping, the little ruby cylinder was aswarm with all-in-step quanta, all identical in wavelength, the characteristically ruby-red light of 6943 Ångstroms. As the coherent swarm came to critical size the reflective power of the thin mirrors was overwhelmed and the cylinder emitted an intense flash of ruby light that lasted perhaps a half-millisecond. Zap!

In the years that followed Maiman's success, all kinds of new facts were learned about lasers. Experimenters pumped up solids other than rubies, elemental gases such as helium and neon and even molecular mixes of gas such as carbon dioxide. By the 1970s it was known that practically any of the elements could be made to "lase," and that laser light was available at wavelengths ranging from about 2,000 Å (two tenths of a micron, at the ultraviolet end of the visible spectrum) to around 130 microns (deep in the infrared).

Not all such lasers performed equally. For example, the pumped-up solids could only emit light in short bursts, whereas the pumped-up gases emitted a continuous beam. The bursts from the pumped-up solids, however, were more powerful than the beams from the gases. Then it turned out that the lasers using mixes of gases — the carbon-dioxide model in particular — could produce a beam of coherent light both continuous and powerful. One of these, at the U.S. Army's Redstone Arsenal in Alabama, developed a power output of nearly nine kilowatts at a wavelength in the infrared range. When focused by means of a parabolic mirror so that its light struck a quarter-inch steel plate, this laser beam could burn a hole through the steel in less than half a minute. Zap! Or, anyway, Zaaaap.

U.S. Army gas laser, 178 feet long, draws its energy from the excitations not of single atoms but of three-atom molecules, in this case carbon dioxide molecules. It discharges a beam of coherent infrared light that, when focused by a concave mirror, will burn a hole through a steel plate one quarter of an inch thick within 10 seconds.

Glow of the tube is due to excitation; the laser beam is invisible.

Star Wars? Well, certainly not terrestrial battles. The Redstone laser was nearly 200 feet long, which is oversize even for heavy artillery, let alone sidearms. Even if smaller lasers could be made as powerful and could be equipped with portable power sources, no enemy target is likely to be decked out with parabolic mirrors that would concentrate the infrared beam.

But what if the beam were prefocused? That's fine, so long as it's not raining or snowing or misty, foggy or smoggy. Any particles in the air scatter the coherent light, leaving the laser beam no more deadly than the incoherent beam of a searchlight or a flashlight.

"Ah," you say, "they ain't no rain in space, so maybe Star Wars after all?" Well, maybe. And surely the military geniuses of the great powers will scratch up enough taxpayers' money to give it a try. This will be at the same time that the great powers' counterweaponeers will have taken a leaf from the book of World War II radar countermeasures and seen to it that military spacecraft are all prepared to dispense "chaff" in the event of attack. Aircraft then (and now) dump aluminum ribbons by the thousands to muddy up the opponent's radar blips. It is not hard to conceive of aluminium-powder aerosols and similar gadgets "armoring" a spacecraft against laser assault. Same thing could be done on the ground, too, in fair weather. Oh, well, it's only taxpayers' money.

Does this mean that the laser as a cutting tool is really a dead issue? Far from it; more applications exist than merely military ones. One cute example is the use of lasers in the technology of "die-drawing." We produce and consume billions of miles of wire every year round the world and all of it is made by drawing ductile metal, such as copper, aluminum, or soft iron or steel, through progressively smaller doughtnut-shaped dies. The harder the die, the longer it will resist erosion and produce wire of the desired diameter. Why not diamond dies, then? Diamond is as hard as they come. Fine, but how do you drill a hole in the diamond? This was a wearying process before the laser came along, but now a simple zap suffices.

Diamonds are a girl's, and often an industrialist's, best friend. How about a physicist's? A problem that plagued experi-

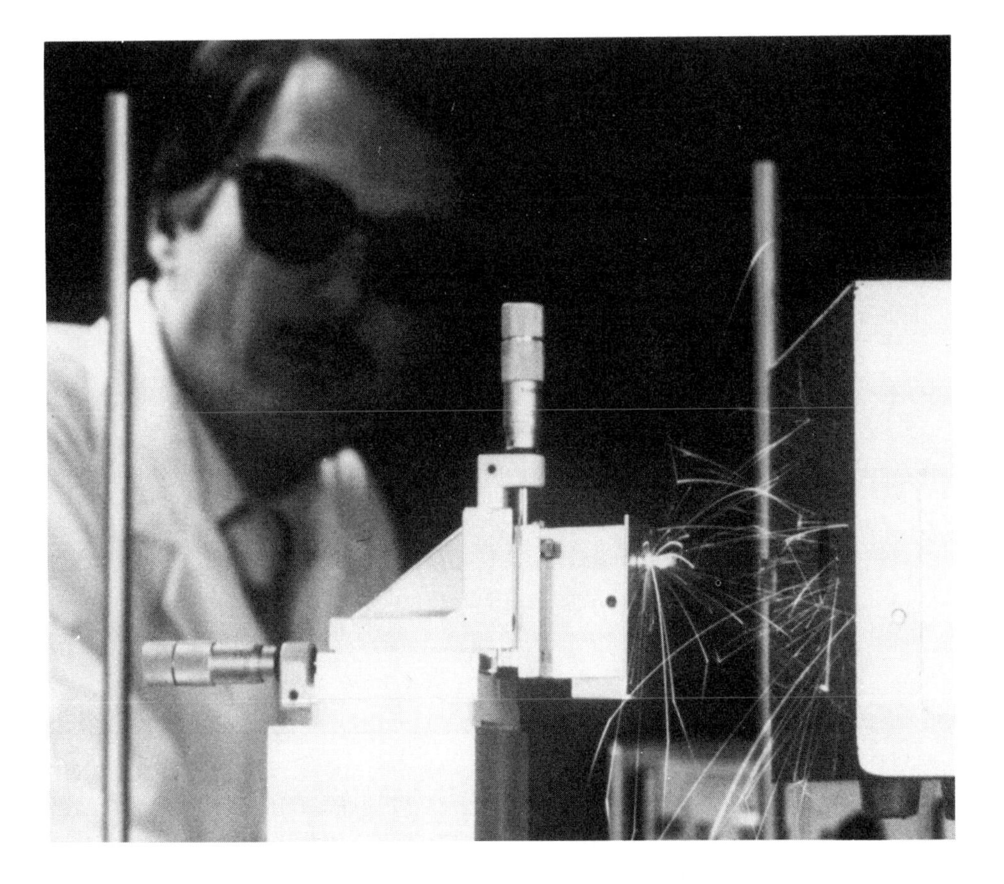

Cutting with coherent light: this laser, being demonstrated in the New Jersey laboratories of the Allied Corporation, is drilling a hole in a stainless steel plate. The light is rendered coherent inside a single crystal of the gemstone alexandrite. This is a tunable laser: the wavelength of the light can be adjusted from 7,000 to 8,200 Å — from near-red on up into the infrared part of the spectrum.

menters at the Carnegie Institution of Washington, D.C., until recently also relates to both diamonds and lasers. The Carnegie workers were trying to create on a laboratory scale the temperatures and pressures that exist in nature only hundreds of miles beneath the earth's crust, near our planet's metal core. Peter Bell, the project leader, solved the problem of applying great pressure by building a kind of vise with diamond jaws. Mineral samples placed between the jaws could be squeezed until the pressure rose to some 25 million pounds per square inch (any greater pressure deformed the diamonds).

But what about higher temperatures? Well, diamonds are transparent. The team rigged a laser so that its coherent beam passed through the diamond and delivered all its energy to the mineral sample. In this way Bell and his colleagues found it possible to simulate the temperature expected at a depth of more than 1,000 miles. Not bad for a laboratory demonstration! (What they learned, by the way, is that the traditional image of the earth's core as a great ball of nickel-iron is certainly mistaken. The iron is there, all right, but either nickel-iron or pure iron would become far denser at core pressures than measurements indicate. Bell has suggested that the iron is mixed with some lighter element, perhaps oxygen.)

Cutting and heating (actually, it's the heating that does the cutting); are these the only peaceful uses for lasers? Certainly not. Various forms of lasing have already been applied to the communications industry (optical cables), the chemical industry (the promotion of chemical reactions) and even in surgery. Many other applications are under development. From my point of view, however, cutting is a good note to end on. Back in Chapter 3, you may remember, I painted the dismal picture of one of our imaginary forebears faced with a potential feast of meat but without any means of butchering the carcass. This made him angry enough to decide it was high time to invent a cutting tool. Later on, as you may also remember, we traced the evolution of man's early flint tools in terms of length of cutting edge per nodule of flint. The choppers that were among the first of these flint tools were named hand axes by the first scholars who were wise enough to recognize them as human artifacts. So now we have traced man's increasing mastery of nature and grasp of energy, cut by cut, as it were, from hand ax to laser. This is a good place on the trail to halt.

BIBLIOGRAPHY

Aigner, Jean S. "Pleistocene Archaeological Remains from South China." *Asian Perspectives* 16:1. Honolulu: 1974.

Asimov, Isaac. *The Intelligent Man's Guide to Science*. New York: Basic Books, 1960.

———. *Realm of Measure*. Boston: Houghton Mifflin, 1960.

Birdsall, Derek, and Cipolla, Carlo M. *The Technology of Man*. New York: Holt, Rinehart & Winston, 1980.

Bordaz, Jacques. *Tools of the Old and New Stone Age*. Photographs by Lee Boltin. Garden City: Doubleday, 1970.

Brodrick, A. Houghton, ed. *Animals in Archaeology*. New York: Praeger, 1972.

Bryant, Lynwood. "The Origin of the Automobile Engine." *Scientific American*, March 1967.

Bulliet, Richard W. *The Camel and the Wheel*. Cambridge, Mass.: Harvard University Press, 1975.

Bureau of Naval Personnel, U.S.N. *Basic Machines*. (NAVPERS 10624-A) Washington, D.C.: 1965.

Charles, J. A., "Early Arsenical Bronzes." *American Journal of Archaeology*, January 1967.

Childe, V. Gordon. "Rotary Motion." *A History of Technology*, edited by Charles Singer et al., 1:9. Oxford: 1954.

———. "Wheeled Vehicles." Ibid., 1:7.

Clark, Grahame. *World Prehistory: a new outline* 2d ed. Cambridge: Cambridge University Press, 1969.

———. *World Prehistory in New Perspective* 3d ed. Cambridge: Cambridge University Press, 1977.

Coles. J. M., et al. "The Use and Character of Wood in Prehistoric Britain and Ireland." *Proceedings of the Prehistoric Society*, London: 1978.

Coon, Carleton S. *The Origin of Races*. New York: Knopf, 1962.

———. *The Story of Man*. New York: Knopf, 1954.

Derry, T. K., and Williams, T. I. *A Short History of Technology*. Oxford: Oxford University Press, 1960.

Drachmann, A. G. *The Mechanical Technology of Greek and Roman Antiquity*. London: Hafner, 1963.

Edwards, I. E. S., et al., eds. *The Cambridge Ancient History*. Plates to vol. 1 and 2. Cambridge: Cambridge University Press, 1977.

Evans, Joan. *Life in Medieval France*. New York: Phaidon, 1957.

Fagan, Brian M. *Men of the Earth*. Boston: Little, Brown, 1974.

Fessenden, Helen. *Fessenden, Builder of Tomorrow*. New York: Coward, McCann, 1940.

Forbes, R. J. "Extracting, Smelting, and Alloying." *A History of Technology*, edited by Charles Singer et al., 1:21. Oxford: 1954.

Fraipont, Julien, and Lohest, Max. "La race humaine de Neanderthal, ou de Constadt, en Belgique." *Bulletins de l'Académie Royale des sciences, des lettres et des beaux-arts de Belgique* 3me Série, 12. Brussels: 1886.

Freeman, Leslie G. Jr. "Chinese Paleolithic Collections." *Paleoanthropology in the Peoples Republic of China*, CSCPR Report no. 4. Washington, D.C.: National Academy of Sciences, 1977.

Garrett, Alfred B. *The Flash of Genius*. Princeton: Van Nostrand, 1963.

Garrod, Dorothy A. E. "Palaeolithic Spear-throwers." *Proceedings of the Prehistoric Society*. London: 1956.

Gimpel, Jean. *The Medieval Machine*. New York: Holt, Rinehart & Winston, 1976.

Gordon, James P. "The Maser." *Scientific American*, December 1958.

Gowlett, J. A. J., et al. "Early archaeological sites, hominid remains and traces of fire from Chesowanja, Kenya." *Nature*, 294 (12 November 1981).

Grigson, Geoffrey. "Old Stone Age: a commentary on photographs by Stevan Celebonovic." *Art and Nature*, New York: Philosophical Library, (undated).

Hadingham, Evan. *Secrets of the Ice Age*. New York: Walker, 1979.

Harlan, Jack R. "The Plants and Animals that Nourish Man." *Scientific American*, September 1976.

Harrison, H. S. "Fire-making, Fuel and Lighting." *A History of Technology*, edited by Charles Singer et al., 1:10. Oxford: 1954.

Hawkes, Jacquetta. *The Atlas of Early Man*. New York: St. Martin's Press, 1976.

———, and Woolley, Leonard. *Prehistory and the Beginnings of Civilization*. (Vol. 1 of UNESCO *History of Mankind*.) New York: Harper & Row, 1963.

Holloway, Ralph L. "The Casts of Fossil Hominid Brains." *Scientific American*, July 1974.

Hommel, Rudolf P. *China at Work.* New York: John Day, 1937.

Iversen, Johannes. "Forest Clearance in the Stone Age." *Scientific American,* March 1956.

Jacobs, Jane. *The Economy of Cities.* New York: Random House, 1969.

Jewkes, John, et al. *The Sources of Invention.* New York: Norton, 1971.

Keeley, Lawrence H. "The Functions of Paleolithic Flint Tools." *Scientific American,* November 1977.

———, and Toth, Nicholas. "Microwear polishes on early stone tools from Koobi Fora, Kenya." *Nature,* 293 (October 8, 1981).

Klein, Richard G. *Ice-Age Hunters of the Ukraine.* Chicago: University of Chicago Press, 1973.

Kurtén, Björn. *Pleistocene Mammals of Europe.* Chicago: Aldine, 1968.

Landels, J. G. *Engineering in the Ancient World.* London: Chatto & Windus, 1978.

Lawrence, Barbara. "Antiquity of Large Dogs in North America." *Journal of the Idaho State University Museum* 11:2, 1968.

Leakey, L. S. B. "Working Stone, Bone, and Wood." *A History of Technology,* edited by Charles Singer et al., 1:6. Oxford: 1954.

Lee, Richard B., and De Vore, Irven, eds. *Man the Hunter.* Chicago: Aldine, 1968.

Littauer, Mary Aiken, and Crouwel, Joost. "The Origin and Diffusion of the Cross-Bar Wheel." *Antiquity,* July 1977.

Major, J. Kenneth. *Animal-powered Engines.* London: Batsford, 1978.

McNeill, William H. *Plagues and People.* New York: Doubleday, 1976.

Oakley, Kenneth. "Skill as a Human Possession." *A History of Technology,* edited by Charles Singer et al., 1:1. Oxford: 1954.

Patterson, Clair C. "Native Copper, Silver and Gold Accessible to Early Metallurgists." *American Antiquity* 36:1, 1971.

Piggott, Stuart. *Ancient Europe.* Edinburgh: Edinburgh University Press, 1965.

Price, Derek de Solla. "Gears from the Greeks." *Transactions of the American Philosophical Society,* NS 64:7, 1974.

Rolt, L. T., and Allen, J. S. *The Steam Engine of Thomas Newcomen.* New York: Neale Watson, 1977.

Schawlow, Arthur L. "Optical Masers." *Scientific American,* June 1961.

Schmandt-Besserat, Denise. "The Earliest Precursor of Writing." *Scientific American,* June 1978.

Shepherd, Walter. *Flint, its origin, properties and uses.* London: Faber & Faber, 1972.

Solecki, Ralph S. *Shanidar*. New York: Knopf, 1971.

Treistman, Judith M. *The Prehistory of China*. New York: Doubleday, 1972.

Usher, Abbott P. *A History of Mechanical Inventions*. Revised. Cambridge, Mass.: Harvard University Press, 1954.

Vitruvius (Marcus Vitruvius Pollio). *The Ten Books of Architecture*. Translated by M. H. Morgan. Cambridge, Mass.: Harvard University Press, 1914.

Watson, H. F. "Undersea Coal Mining in North East England." *Proceedings of the Royal Society for the Encouragement of Arts, Manufactures and Commerce*, January 1979.

Wendorf, Fred, and Schild, Romuald. "The Earliest Food Producers." *Archaeology* 34:5, Sept./Oct. 1981.

Wheeler, Mortimer. *Civilizations of the Indus Valley and beyond*. New York: McGraw-Hill, 1966.

White, K. D. *Agricultural Implements of the Roman World*. Cambridge: Cambridge University Press, 1967.

White, Lynn Jr. *Medieval Technology and Social Change*. Oxford: Clarendon Press, 1962.

———. "Medieval Uses of Air." *Scientific American*, August 1970.

Whitehouse, David, and Ruth. *Archaeological Atlas of the World*. London: Thames & Hudson, 1975.

Wolpoff, Milford H. *Paleoanthropology*. New York: Knopf, 1980.

Woodcroft, Bennet, translator. *Pneumatics* (The *Pneumatics* of Hero of Alexandria), London: 1851. Facsimile edition, London: Macdonald, 1971.

Zhang Sen-shui (Chang San-sui). "Fifty Years of Paleoanthropology in China." *China Reconstructs* 29:4. Beijing (Peking): April 1980.

INDEX

In the following SUBJECT index entries that commence in boldface type indicate an illustration or a citation in the caption accompanying an illustration. A GEOGRAPHICAL index begins on p. 307.

GEOGRAPHICAL INDEX